METALWORKING
DOING IT B

Machining

Welding

Fabricating

Tom Lipton

INDUSTRIAL PRESS INC.

A full catalog record for this book is available from the Library of Congress.
ISBN 978-0-8311-3476-1

Lipton, Tom.
  Metalworking doing it better: machining, welding, and fabricating/Tom Lipton.

  ISBN 978-0-8311-3476-1 (softcover)
    1. Metal-work.   I.   2014

Industrial Press, Inc.
32 Haviland Street, Unit 2C
South Norwalk, CT 06854

Sponsoring Editor: John Carleo
Developmental Editor: Robert Weinstein
Interior Text and Cover Design: Janet Romano
Composition: Thomson Digital

ACKNOWLEDGMENTS

When I completed work on my first book, *Metalworking: Sink or Swim*, I included the following acknowledgments, which still hold true today.

This book might never have existed had it not been for a few important people. I would like to take a moment here to thank them properly.

Without my wife Sargamo, I don't think this book would have ever been finished. She was able to cheer me on at key points and breathe some life back into me. As a fellow metalworker, she can read this material and understand it fully. We met in a welding shop 25 years ago and, for some unknown reason, she has not cut me loose yet. We have the dubious and unique honor of having the worst fight of our marriage over a pair of really nice C-clamps at the flea market. She can bring home the bacon as well as weld me under the table with one hand tied behind her back. In addition to her metalworking skills, she doggedly tried to improve my grammar and punctuation in micrometer-like steps.

I would like to thank all the metalworkers who have gone before me, on whose shoulders I am standing. Looking ahead to the future, I don't like what I see for the skilled trades. I am doing everything I can to make sure nothing dies that shouldn't. I have learned from so many people it would be a book of its own to thank everyone properly.

Chris Owen may not be what I would call a career metalworker, but I still owe him a little
credit. He had the dubious honor of letting the book-in-progress out of the bag at WESTEC 2006 to an unexpected warm reception. By the way, Chris, you still owe me a lunch.

I believe that my parents had a strong hand in shaping my life as a metalworker — from the day when I was nine in the basement learning how to weld with my dad to my mom fronting the money for my first welding machine. How could I fail with support like that?

Thanks to all!

Tom Lipton
August 2008

Now as I complete work on *Metalworking: Doing It Better*, I want to say thanks to a few important people who helped me with this second volume on my favorite subject, metalworking. Without support and encouragement from my publisher Industrial Press and their professional editing, this volume would be less perfect than it is now. Thank you to Robert Weinstein for doing the heavy lifting and to John Carleo for encouraging me to take on this project in the first place. Another special thanks goes out to the amazing Janet Romano for wading through a truckload of pictures of strange things and arranging them into a meaningful coherent work. I had the easy part. All I had to do is do what I love — which is the metalworking part. These dedicated folks really insisted and worked hard to make sure that it would be a really good book. I hope you agree.

My wife deserves another special thanks for taking yet another bullet for the team. She still has not cut me loose — she put up with me huddled behind the computer whittling away when she wanted to watch *Breaking Bad* or *True Blood* episodes. I owe her big time on this one.

I'm sure there are a few more folks who I missed and who deserve some acknowledgment for the parts they played in getting this book to you; I apologize in advance because it is definitely not intentional.

Sincerely,

Tom Lipton
August 2013

TABLE OF CONTENTS

Acknowledgments   v

Chapter 1   Diving In   1
  1.1   Welcome to Doing It Better   1
  1.2   Personal Learning Attitude   3
  1.3   Shop Environment   6
  1.4   What's a Journeyman Anyway?   7
  1.5   Thursday Nights   8
  1.6   Format   10

Chapter 2   Brain Food   13
  2.1   Communication   14
  2.2   Drawing and Sketch   17
  2.3   Minimizing Screw-Ups   24
  2.4   Accuracy   25
  2.5   Speed   25
  2.6   Shop Math   26
  2.7   Mass, Volume, and Area   30
  2.8   Angles and Shop Trigonometry   30
  2.9   The Metric System   33
  2.10   Computers and the
         Metalworker   35
  2.11   Dumb and Dumber   37
  2.12   Want to Make a Million
         Dollars?   40

Chapter 3   Bean Counters Lounge   43
  3.1   Engineers and Metalworkers   44
  3.2   Shop Talk   45
  3.3   Dimensioning   48
  3.4   Other Tips   49

Chapter 4   Setting Up Your Shop   51
  4.1   Floors   52
  4.2   Light   54
  4.3   Food Areas   55
  4.4   Heating and Cooling   55
  4.5   Workbenches and Tables   56
  4.6   Air Supply   58
  4.7   Raw Material Storage and
         Handling   62
  4.8   Material Identification and
         Characteristics   65
  4.9   Safety Equipment   76
  4.10   Tool Crib   79
  4.11   Bench Work   83
  4.12   Filing   86
  4.13   Saws and Sawing   93
  4.14   Rigging and Lifting   116

Chapter 5   Manual Lathe   125
  5.1   Learning to Love the Lathe   126
  5.2   Getting Started with the
         Manual Lathe   127
  5.3   Step Turning   143
  5.4   Threading in the Manual
         Lathe   144
  5.5   Multiple Start Threads   147

**Chapter 6   Manual Milling
               Machine**   153
  6.1   Bridgeport Mills   154
  6.2   Suggested Improvements   155
  6.3   Spherical Surface Generation   174

Chapter 7 CNC Mill   181
  7.1 Working with CNC
       Equipment   182
  7.2 CNC Mill   184

Chapter 8 CNC Lathe   207
  8.1 CNC Lathe Programming   209
  8.2 CNC Lathe Part Making   212

Chapter 9 The Welding Shop   219
  9.1 Getting Started   221
  9.2 Layout Work   222
  9.3 Some of My Favorite
       Hand Tools   233
  9.4 Welding Tables   235
  9.5 Brake Bumping   246

**Chapter 10 The Lost Art of Flame
Straightening   259**
  10.1 Limitations   262
  10.2 How Flame Straightening
        Works   262
  10.3 Heat Input   263
  10.4 Mapping   264
  10.5 Applying the Correction   266
  10.6 Straightening Shafts
        and Tubes   268
  10.7 Special Applications of
        Heat Shrinking   270
  10.8 Correcting Weldments   271

Chapter 11 Sheet Metal Shop   273
  11.1 Layout Work   275
  11.2 Blank Length Calculations 276

  11.3 Patterns   283
  11.4 The "Yank" Method   284
  11.5 Forming and Layout of Cones 297
  11.6 Tanks and Baffles   301

**Chapter 12 The Abrasion
Department   303**
  12.1 Sanding, Grinding,
        and Abrading   303
  12.2 The Good, the Bad, and
        the Ugly   317
  12.3 Radius Grinding   319

Chapter 13 The Junk Drawer   321
  13.1 Miscellaneous Tricks
        Without a Home   321
  13.2 Ideas for the Shop Floor 325

Closing Thoughts   331

**Appendix A: Squaring Blocks
without a Tool Change   333**

**Appendix B: Recommended
Reading List   339**

Index   341

Diving In

1.1.   Welcome to Doing It Better
1.2   Personal Learning Attitude
1.3   Shop Environment
1.4   What's a Journeyman Anyway?
1.5   Thursday Nights
1.6   Format

1.1 Welcome to Doing It Better

I always wanted to write a book. I started more than my fair share actually, but for numerous mostly lame reasons never was able to finish any. Then a few years ago, *Metalworking: Sink or Swim* came together. I've been pleased with the result and am very happy now to be moving forward with other book projects like this one. Whereas my first book was aimed primarily at machinists, welders, and fabricators who work in larger metalworking shops, this book as aimed toward small scale DIY (do it yourself) operations, including those you might find in maker shops and small business.

Career Metalworker is probably the best way to describe me. As I work on this manuscript, half a century has passed under my bridge since I was hatched and I still love metalwork. I may not have loved all the jobs I have done or all the places I have worked, but I have always loved the trade. My parents might have had a small clue I was destined for the trades after my mom gave me the serrated saw off an aluminum foil box, which I put to good use sawing up the arms of her nice dining room chairs.

Maybe a more accurate description of me would be that I love the skilled trades. I can appreciate and be humbled by the violinmaker, the plumber, and the tugboat captain. The act of building something is deeply satisfying and difficult to explain to people outside the trades. Wherever humans pick up tools and work with materials, machines, and a skilled hand, this is where I want to be.

The only caveat is that it must be done well. This appreciation for attention to detail was drummed into my head by some of the old guys, more by osmosis and boot in the rear than any direct action on my part. I always felt that I let them down if I did a bad job or something didn't quite come out the way they wanted it. Most of the time, however, they didn't even have to say anything. You knew from the look on their faces or a callused hand sliding over the offending detail that you had somehow failed slightly. The answer for me was to try harder the next time and learn from the experience.

Call me strange, but I love the sight, sounds, and smells of a working shop (see Figure 1-1). Each has its own distinct flavor and heartbeat. The smell of hot metal and cutting oil brings certain memories out in clear relief. Almost any welder can smell a piece of paper burning halfway across the shop in the middle of cutting steel plate with a torch, sniffing the air like a bloodhound looking for the start of a fire. We can tell which shop the boss sent the grinding work out to by the smell of their cutting oil.

These shops we work in get into our blood in more ways than you know. The squeal of a tortured cutting tool, the clank of a pair of vise-grips opening, or the sound of a tack weld breaking is as recognizable as your own name called by your mother in a noisy room.

My first experiences in metal working started with welding. My father taught me how to stick weld when I was nine years old in the dark basement of our house in Berkeley. Like the sailors who get the open ocean into their blood, I can say this is the moment I was infected by a fascination with metalworking or, at least at that point, welding.

In school I took machine shop and welding and never looked back. Somebody I thought was smart told me back then that having two different sets of skills was a valuable asset. They could not have been more right and it has served me well for a long time. The real message was never stop learning about your trade.

Young people just entering the trade are encouraged to stick with it and get through the tough beginning years. Things will still need to be built from metals and the trade needs new talent to advance. Be versatile and don't shy away from the tough jobs—you will be rewarded with a lifetime of support and hopefully enjoyment of a job well done. These first few years are the character building years where you "Pay your Dues" and learn an appreciation for all the aspects of your trade.

Figure 1–1 A small, well set up hobby shop.

1.2 Personal Learning Attitude

Your attitude is one of the key ingredients to success in any field, not just metalworking. Without a positive and persistent attitude, you might as well just go sit in front of the TV and bathe yourself in some nice cable programming. The power of learning and dogged persistence cannot be overstated. My wife and I call it burning rod.

You have to burn rod and put your time in to learn how to weld or become skilled in any trade (see Figure 1-2). In my experience, most people don't learn on the first rod they burn or the first thread they cut. Winners do what losers are unwilling to do.

We are in the middle of a unique time in history. The ability to share new ideas, information, and old skills will never be better. This critical time balances between the new guard and the old. On one side, we have access to technology for sharing huge amounts of detailed information across thousands of miles and time zones in the blink of an eye. On the other side, we still have access to

the people and knowledge whose shoulders we are standing on and who form the foundations of our trades. This combination of factors has not been the case throughout history.

Not that long ago people never traveled more than ten miles from the town where they were born. Ten miles represents the distance you could walk and return home in one day. Anything outside that radius might as well have been imaginary.

Right now in our time, I can move my finger and click two or three times and look at the surface of another planet in our solar system. That to me is truly amazing. Now you can learn or teach all the way around the planet. Borders and time zones have no real meaning now for the learning process. You will either be in this wave of learning, or be left behind and fossilized by it.

This book is about learning new, and advancing current metalworking skills. The trades have been very good to me. Part of the requirement the trade imposes is to pass on knowledge and skills to those willing to learn. We have all stood on the shoulders of the people we have learned from; we owe at least the payment of passing the skills on. Each generation should push the boundaries of their art to the next higher levels. I thank the people I have learned from because without them I would still be trying to figure out how to smelt iron.

Your attitude toward learning and your skills are your protection in modern times. No longer can you rely on having a good job for life working for a stable company. Entire industries are being created or becoming obsolete on a daily basis. Modern skilled tradesmen have to constantly adapt and add skills to their toolkit to keep up with the pace of industry and the modern global electronic economy. Your skills must not stay static. Learn everything you can about everything. Sink, swim, or get the heck out of the way.

Figure 1–2 Burning rod.

The advancement of any craft depends on new experiences and new people with sometimes wild and exaggerated ideas who push the boundaries of current knowledge or accepted practice. This is one of the character traits that built America. As new techniques, ideas, and materials become available, postpone or suspend judgment about them. Look at how they might be applied instead of dismissing them. An open environment where every person and new idea has worth—without concern for criticism or dismissal—is key to success. Truth and accuracy in knowledge and information, and the destruction of myth and misinformation, are required to further the art. Speak the truth; walk a reasonably toleranced line.

It never ceases to amaze me how some people memorize sports trivia, yet more often than not they are the same people who ask how to run a particular machine in the shop they have been walking past for ten years. Instead of investing time and effort to improve themselves, they choose to invest in a big screen TV or an F-350 4X4 turbo diesel to haul groceries. These are some of the same folks who will ask to borrow your tape measure because they don't have one.

Typically, these same people never have enough of anything. More money, more beer, more toys, more horsepower. They don't correlate that their skills = value = profit = wages. Notice wages comes after profit. The companies we work for or start and run must be profitable or they cease to exist. As skilled tradesman, we have the responsibility to use our skills to make sure our companies survive and prosper.

Every company is built on people. Machines and materials are commodities that can be bought, sold, and traded any day of the week. Great people are grown, cultivated, protected, and nurtured. In exchange for this, they give back loyalty, dedication, innovation, and hard work.

It's called a trade for a reason. Don't misunderstand what I am saying here. I enjoy leisure time as much as the next guy, but I also love my work and would be doing the same thing even if somebody didn't conveniently pay me to do it.

You cannot learn skilled trades by reading a book, even this book, period. You can get an understanding of the technical issues and the tools involved, but true skill comes from hands-on practice. Anybody can learn some metalworking trivia and talk a good technical line. Just like a good salesman, they can sell themselves like a shiny new car. But, there is nowhere to hide out in the shop when the rubber hits the road. You either go up in smoke or gain traction. An imposter stands out like a cow-pie at a croquet match to someone skilled in the trade.

An interesting example of this comes from a story a friend told me about passing through international customs one time. The customs inspector asked him what kind of work he did, to which he replied that he was a machinist and worked with metal. The next question from the customs inspector was, "Let me see your hands" This is pretty telling, that you can judge a person's validity by looking at their hands (see Figure 1-3). All I can say is 'good luck' if you're an imposter!

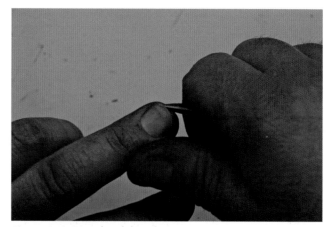

Figure 1–3 Meathook hands.

Figure 1–4 Charlie Blessing and Doug Duane, Master Tool-maker and Sheetmetal Man.

Skills are like calluses; the faster you try to go, the easier it is to get a blister. The slow, steady approach builds skills and calluses for a lifetime of learning and rewards. If you think you can read a book or take a pill and miraculously emerge a seasoned veteran, you are mistaken. It takes years to hardwire the necessary muscle memory to perform some of these operations, but once you have it is obvious to others in the trade. It's the little things that give away the masters—like the way a sheetmetal man flips his wrist to fan a cushion of air between sheets of metal or the little quick head nod of an experienced welder putting their hood down. You can't learn these things overnight.

I know for myself, and I'm pretty sure it's the same for most, the way I learned is in little gems and nuggets, during the process of making lots of mistakes. Slowly you gather these small parts together of the really big puzzle. That why old geezers are so darned smart; they have been picking up pieces of the puzzle for a long time. They've have had a lot time to gather and polish their nuggets. The trick is to write or pass them down before your memory starts to fail. That's what I'm trying to do with these nuggets.

Somehow I have been lucky to develop good relationships with several great teachers (see Figures 1-4 and 1-5). I sometimes feel pathetic and puny next to their skills.

The only way I could ever hope to surpass them is if they die and give me a chance to catch up a little. That's exactly what's happening.

My old teachers and workmates are dying off. When they die, they become static points. All their amassed skills stop growing and start to dissipate until they disappear forever or, worse, have to be learned again. By writing down and documenting as many as I can remember, I can preserve them for future tradesmen. So do your part for the trade. Take some old geezers under your wing and learn something from them. The baby boomers will be retiring in droves in the next several years, taking all their wonderful hard-earned knowledge with them right out the door. I have tools that were given to me from some of the most influential people I learned from. Every time I pick up one of these tools and use it, a flood of memories comes back along with a deep appreciation and feeling of humility. I can still hear them telling me to be careful and not screw up that nice tool I gave you.

In ten thousand years, modern humans will have forgotten how to read the ancient language this book is written in. But I am willing to bet they will still use metals and need to fabricate them into useful articles. I'm sure the methods will be different. However, I am confident they will have their roots in the things we know now and have learned from the people that went before us.

Figure 1–5 Fred Van Bebber, Master Machinist

1.3 Shop Environment

All work and no play make for a pretty dull shop. Working in a shop with a bunch of other people is somewhat like a marriage. There are good, bad, and really funny days. Just like a family, there are members you get along with and others that you don't. You spend more waking time at work with your workmates than you do with your mate or immediate family in a given work week. If you can't have a little fun and get along, it makes for a pretty miserable time. I have purposely included an attempt at humor in some of the descriptions and pictures.

Over the years I have worked in many shops, some large and some small. Overall I prefer the small shop dynamic. The flavor of a shop is created by the people working there. Shops can foster and nurture the learning and skill building attitude or they can undermine and destroy it. It is a choice.

One of the greatest gifts I have been given is the thirst for skill growth. This sounds simple enough, but is much harder to do in practice. If you were to ask anybody if they support skill growth or learning, what do you think they would say? In all likelihood they will agree and say "yes." The only way to truly judge this is by actions. How do you support this by action? Humans learn best by doing things. In particular, things they are interested in. When you are interested in something, the learning is almost effortless.

Have you ever been surprised by how different and incomplete the descriptions of tasks and operations you read about in a book were when you actually tried them out for real? If you were lucky enough to learn some of your craft in an apprenticeship program, then you know what I mean. This balances the need for some theoretical work with a healthy dose of doing things in the shop.

Here are two scenarios to think about.

You're sitting in a classroom listening to the teacher talk about how long it takes a train leaving Chicago to get to New York if it's traveling east at 60 mph and another train leaves New York, yada, yada,........ This is what I call linear or structured learning, otherwise known as lacking moisture and inductive of sleep. Almost all schools and academic institutions use this method. They start at the beginning and move in a deliberate step-by-step fashion until they get to the end. Some people thrive in this type of cranial learning environment. Some of these folks end up as engineers or scientists along with business managers. Normally, they are politely called white-collar workers by shop folks. These are the same people who write textbooks and decide how to train and improve the other type of workers. You know, the ones with blue collars and dirty jeans.

Most trades people have learned their skills in a much different way. In fact, some may have gone into the trades because they didn't like the structured linear method. The main method of learning skilled trades is the direct hands-on method. In my experience, the learning bounces around more and is definitely less linear than in a school environment. Typically whatever you were working on was the subject of the lesson that day. Most of us blue-necks have come to be where we are by this bouncing around method.

1.4 What's a Journeyman Anyway?

There are many names for the seasoned tradesman. Typically, these professions were dominated by men. The names reflect that fact. It is not intended to be derogatory or sexist—only historically accurate.

Journeyman, Apprentice, Tradesman, Master, Rookie, and Craftsman are all names associated with different skill levels related to the skilled trades. Most trades have no published standards or colorful martial arts belts given out to indicate specifically what it means to be an apprentice or the tenth-degree grand master.

In my mind, it is not the number of years served in a trade but a question of ability. All too often people are given a title just because they have a certain number of years at the bench. I have seen 40-year veterans who stopped learning after their second year and became miserable static points.

I have also seen 4-year apprentices who were literally sponges starved for information, easily exceeding their static counterparts. There lies the problem. How do you measure ability? The definition of ability is different for each trade and cannot be measured merely by the passage of time. The only answer is for other top people in the specific trade to establish and make the ability assessments.

My definition of the top meat eaters of the skilled trade food chain goes like this. It's the people who have enough skill and experience to draw on in order to take on any problem that comes up in their trade. They may not know exactly how to tackle every job, but they have the experience and acuity to chip away using their skills and experiences to get almost any job done well. Journeyman cavemen can catch, cook, and clean their dinner as well as make a more efficient spear from one of the leg bones—and then go out and do it again and again day in and day out.

The French apprenticeship association (Les Compagnons du Devoir du Tour de France) has a system of skill measurement that seems perfect. It has withstood the test of time, 400 years and counting. After a certain number of years in a particular trade, you must submit a project for your masterwork. No term paper, no book report or thesis, but some real down-and-dirty work. I guess they figure if you hang around for four or five years, you will have at least learned which end of the hammer goes down, but they still want you to prove it.

This project proposal is reviewed by a panel of masters in that trade. If the project is judged difficult enough to demonstrate a high level of skill and competency, then you're off and running. No special time off or preferred treatment is allowed. The project must be completed along with all the other responsibilities the student has. Gee, it's kind of like the real world—pressure included.

The completed masterwork is presented and judged by a panel of masters in the particular trade. If it passes scrutiny, the applicant is awarded their master card (pun intended). In the case of the Les Compagnons, it is a cane or staff with the colors of their chosen trade on it. If the project is not deemed difficult enough or will not demonstrate the proper combination of skills, it is rejected. The applicant must then submit a new project or modify the original project. If you don't complete the project, you get to stay mucking about in the lower levels forever. If students do a lousy job, then I imagine they have to wait to try again.

1.5 Thursday Nights

Fairly early in my career, I was very lucky to work at a shop that supported hands on skill growth and learning in a unique way. I don't think they realized what they were doing and it certainly was not intended for the purpose of training. I only realized it many years later when I was in the position to implement a similar setup.

The shop allowed us to work on small personal projects using the company facilities and resources (see Figure 1-6).

It sounds pretty dumb and simple, but there is quite a bit here. One of the old guys I worked with at this company was employee number 001. He was the first employee and for many years he was the shop foreman until a serious industrial accident sidelined him. He started the tradition of what I call Thursday nights. This was the special time set aside for guys to work on their own projects. It was only one day a week but it was sanctioned, albeit weakly, by the company. We were allowed to use company equipment and minor materials and build almost whatever we wanted. This special time was set aside so the inevitable "G-Jobs," or personal projects, would not be done on company time. He used to say,

"Every man has a little bracket in their life."

Thursday nights had a much more important effect.

What happened was the guys would build things. More important, they built things that they were interested in. They were gaining hands-on experience in the best possible way, by doing. You could build almost anything including things you had never built before. Things the shop foreman would never assign to you because you didn't have the skill or experience. Obviously there were some limits on what you could put together.

The most extreme example I know of took place over a fourteen-year period when a friend built a 48- foot sailboat. He built the boat in his backyard but almost all the fittings and bits crossed his workbench at one time or another. Another guy built a stainless steel hot tub. The list goes on and on.

Well, I say, if you have the experience, who cares how you came by it? So by doing things and trying things you had never done on Thursday night on your own time, you built up your skills. I look back and the most successful people at that company were the ones that were there every Thursday night chunking away on their own projects and "brackets." Most success, either career or financial, can be traced to proactive learning behavior.

Figure 1–6 Here's a helve hammer I built on Thursday nights.

Basic Thursday night rules

- Two or more people minimum working together. No one works alone. Somebody is responsible for locking up.

- You cannot disturb any company work in progress. That is, unless you finish it. I have seen quite a few company jobs completed on personal time just so a machine could be used for a minute.

- You cannot run a business out of your employer's shop. That means no work for money. Trades and barter are okay, but no visitors. The only exception to the no visitors rule is if they bring food.

- Ask permission to use any company materials and pay for any major materials. The foreman has the say so on the amounts and types.

- Do not use company floor space to store personal works in progress. It goes home every night.

- Clean up your tracks. Ideally the process should be invisible. Nobody will complain if an occasional six-pack of sodas shows up in the lunch room fridge. Show your appreciation by example, not talk.

The rewards are the students gaining skill and position while the company gains skills, versatility, loyalty, and dedication. What more could you ask for in a fair trade?

Of course, you can have a classroom and books along with written tests. But the true measure is: can they do the work? "They passed the written test but they can't find the start switch" is the all too common result of book learning. How many drivers' licenses are handed out by just passing the written test and a minimal hands-on demonstration of driving skills? I see the results of that method every day on the highway during the sleepy morning migration and the angry afternoon free for all.

You really have to push the pedals and get dirty to learn this work. You can't learn skydiving or how to pull nine G's in a fighter plane without getting off the ground. Another example of this idea shows up clearly in a hiring situation. Suppose you have your choice between two candidates—one who had practical experience and time in the shop doing a particular operation, and another who had taken a class and passed a test but had not been in the shop. Which candidate would you favor?

With all that said let's get to the good part. Good luck and never stop learning.

"The price of failure is only knowledge."

Figures 1-7, 1-8, and 1-9 show things I've built over the years with my Thursday night program just for the love of building things.

This kind of program provides the fastest, most effective way I know to gain important skills. I support this program in the shop I manage, with excellent results.

The program also delivers a super-positive message from the company to the employees, supporting them by trust and positive action. The company invests the materials and machinery along with space and the students invest the time.

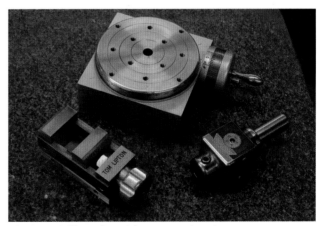

Figure 1–7 Thursday night personal projects.

Figure 1–8 Thursday night pyramid rolls.

Figure 1–9 Thursday night English wheel.

1.6 Format

Finding the best format has been one of the hardest parts of writing this book and has contributed to the failure of previous efforts. How could I present this kind of information to readers in a way that keeps them interested, yet is not so bloated and long winded that the true gems of information become lost.

In the end, I decided to use a format that reinforces the way this kind of information gets into our heads in the first place. Call it a recipe book with each item standing on its own like a good meal and becoming a part of the whole experience. Lots of pictures contribute to a magazine-like format that can be snacked on in little bites any time you feel like opening the book.

Not all the ideas in this book will be of use right away. Your mind and situation have to open and ready to receive. Others bits will be immediately useful whereas some may never be. Heck, I'm pretty sure I will even be accused of "copying" ideas. Remember folks, the knowledge belongs to the trade, not to any individual, me included. I will be the first one to admit that I learned many of these things from other people and by keeping my ears open and my mouth shut.

I hope this book will be something to refer back to and even add recipes of your own to become a kind of larger metalworking cookbook. Many things written here are directly related to my own personal experiences (see Figure 1-10). That does not automatically make them right for everybody. Use common sense and decide for yourself what makes the most sense to you.

If you are offended by anything I write here for whatever reason, please try to get a grip on reality; otherwise you will never survive in a real shop. In fact, let me say that anybody who is offended by this material is a prime candidate for a life of torture and harassment in pretty much any metalworking shop. Don't let the informal style of my writing fool you. Trust me; there are some great nuggets in here. Potentially even one of the suggestions you find here could repay the investment in this book a thousand times over. I admit this material doesn't read like a Tom Clancy techno-thriller, but if you enjoy your trade, you should find some useful information in here and I almost guarantee at least a chuckle or two.

There are many excellent, well-written books on the basics of these subjects. Some very good ones are listed in the bibliography and recommended reading list. This book is designed intentionally for metalworkers who already have a solid background in the basics, like righty tighty lefty loosey. I make the assumption that the readers already know whether their rear ends have been drilled, punched, bored, or reamed….

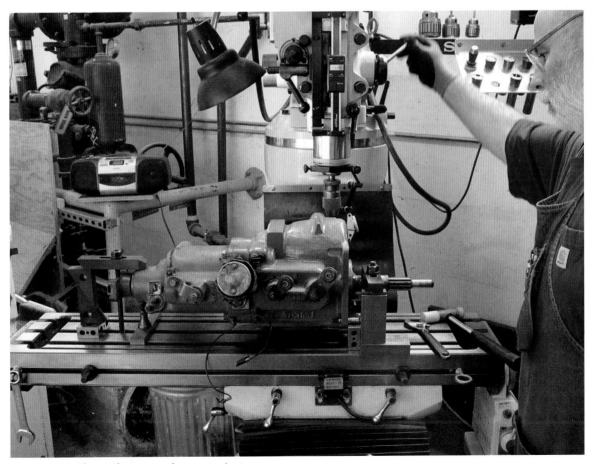

Figure 1–10 The author at work on a project.

2.1   Communication
2.2   Drawing and Sketching
2.3   Minimizing Screw-Ups
2.4   Accuracy
2.5   Speed
2.6   Shop Math
2.7   Mass, Volume, and Area
2.8   Angles and Shop Trigonometry
2.9   The Metric System
2.10   Computers and the Metalworker
2.11   Dumb and Dumber
2.12   Want to Make a Million Dollars?

Brain Food

Not all the skills we need in the metalworking shop involve the hands. Our brains are the most powerful tools we have at our disposal if we will just pry them open a little. Many times we can think our way out of a problem if we give ourselves half a chance. Improvement in thinking starts with admitting you don't know something.

Have you ever had a serious discussion with someone you admired and thought was extremely intelligent? If you pay attention, when you talk to people like that you will notice that you end up explaining many things in great detail about your areas of expertise to them. Why is that? I think it's because for years they have been able to admit they don't understand something, which opened their door to learning. They are able to ask even the most mundane questions because they readily admit they don't know something.

As soon as you say "I don't understand," you open the door. You can't put money inside a locked safe; the door has to be opened and the combination starts with "I don't understand."

13

2.1 Communication

Whether you work in a large or small metalworking shop, communication is an important tool of the trade. Even if you are an occasional hobbyist, with a backyard shop, communication helps you keep on top of projects. Effective communication is quick, clean, accurate, and—most important—to the point. A good engineering drawing has all these elements. Have you ever noticed that the best drawings have no questions? Effective written and visual communication can be considered a form of language. If you don't speak the language, you are condemned to get your information from people who do. A stranger in a strange land is not in a friendly place.

A short note or sketch is worth its weight in gold compared to even the most detailed verbal information (Figures 2-1 and 2-2). If you just counted the parts in a box, take a second and leave a note for the next person so they don't have to re-count (Figure 2-3). Often the note is for yourself, and saves you time and effort when you return to a job. The note is a physical representation of a single thought that all you have to do is see it and you instantly remember.

When you have to turn over a job to someone else, leave a brief note, even if it's just to say the

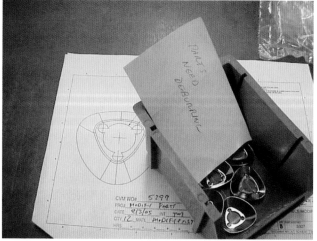

Figure 2–2 Leave a note for the next person – including yourself!

job is finished. Some of these things are plain old manners that your mother should have taught you already.

If you must receive information in oral form, do yourself a huge favor and write it down immediately. A few quickly taken notes may just make the difference in doing a job correctly. Documentation wins over memory when the chips are down and hats are being handed out for rear ends.

A simple elevation of your trade would be just to learn how to spell, or at least use words that you can spell correctly. I'm still working on this one….

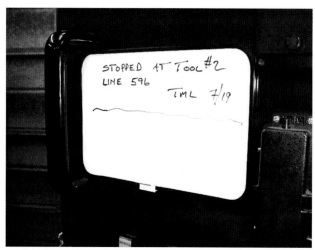

Figure 2–1 A simple note prevents confusion.

Figure 2–3 Keep records of what's been completed.

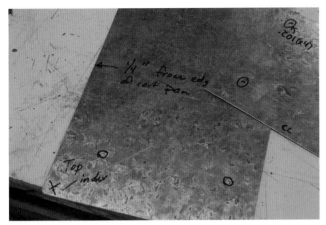

Figure 2–4 A physical template.

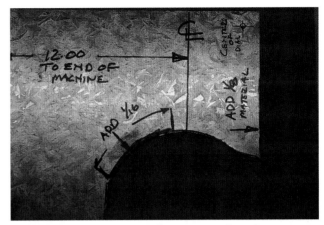

Figure 2–5 Notes on a template communicate important information.

Make accurate physical templates and take copious written notes in the field (Figures 2-4 and 2-5). Don't alter the information that a template made in the field is communicating. The template made in the field and your written notes are gold when you're back at the bench. Typically it is a real pain to get a second chance to take field measurements. Therefore, when you do have the chance, overdo it. Pretend a lawyer will be looking at your templates and field notes and asking those kinds of questions only a lawyer can ask.

An inexpensive digital camera can collect information faster than an army of draftsmen or court stenographers. These are so inexpensive and easy to use that there is no excuse why every shop doesn't have one. Take an extra few shots from slightly different angles. You never know which photo will become the real lifesaver. My rule is that when I think I have enough pictures, I take ten more for good measure. It's the pits when you get back to find out some photos are out of focus or the flash washed out the critical detail you needed. This problem became painfully obvious when I started my first book. Digital pictures are basically free data. Most of the time, they are never printed, but can provide invaluable documentation.

Mark or otherwise identify bad parts as soon as it is known they are bad (Figure 2-6). They may still have some other use, but the parts must be unmistakable and easily identifiable from the good ones. This labeling communicates to others without interpretation or question what the status of a part is. If you don't mark them immediately, then others who pick up the part will have to figure it out themselves, driving the cost of the scrap part even higher.

Figure 2–6 Identify bad parts.

Figure 2–7 Roughing with a reliable measuring tool.

Do your roughing with the most foolproof measuring tool you have (Figure 2-7). As you refine your dimensions, change to the appropriate tools for the level of accuracy required: scale, calipers, and then micrometers. It's hard but not impossible to screw up a 12-inch dimension when using a 12-inch ruler to take the measurement.

Machinists don't make mistakes in one-thousandth increments. They miss by .100 or 1.00. The classic mistake is one revolution of the micrometer barrel, or .025. I've seen some gut wrenching mistakes made because of one turn of the barrel.

Welders and fabricators typically miss in one-inch increments. When you make your first measurement, be sure to force yourself to read the actual number, then the fractional part. You would be surprised how often this mistake happens.

Leave a note for yourself if you need to remember something (Figure 2-8). In a pinch, I may write on my hand. It's easy to forget something if you have lots to remember, so write as many down as is practical.

Another trick I use for something really important to remember is to leave my car keys with what I'm supposed to remember. You won't get very far without realizing that you are supposed to remember something.

Mark your drops and scraps early (Figure 2-9). If something is worth saving, you had best know what it is.

Ask questions early and make sure you understand the answers. There is nothing worse than missing something and having to ask again and again. It's really easy to say "I don't understand" or "I'm not with you" early on. It's much harder to ask later after you were supposed to know.

Figure 2–8 You don't need paper to write a reminder!

Figure 2–9 Labeling saves time later.

2.2 Drawing and Sketching

Drawing is the common language of our metalworking professions. Except for minor differences in conventions, two metalworkers in different countries should be able to exchange drawings and be able to interpret the specifications. This ability to communicate across verbal language barriers is extremely powerful. We can drive a car in a foreign country because of the common factors related to driving. Drawing and sketching are the same. Making up your own drawing conventions is just like really bad driving—everybody hates it.

Use 8-1/2 × 11 sketch paper. A bad sketch on small paper is a worse sketch. Those little free pads the metal suppliers give you are meant to make you screw up your work with bad small drawings so that you have to buy even more metal….

Draw large and fill up the full sheet. Tiny sketches are like tiny brains: who wants one?

You can't borrow a sheet of paper. Have you ever had one returned to you?

Press hard with your writing instrument. Faint lines and text are for accountants, not metalworkers. I like to see the little blivits where the lead actually broke because you pressed so hard. Everything shows up when it goes through the copier if you use a heavy hand. Anybody who uses a pencil with a .5mm lead does not press hard enough.

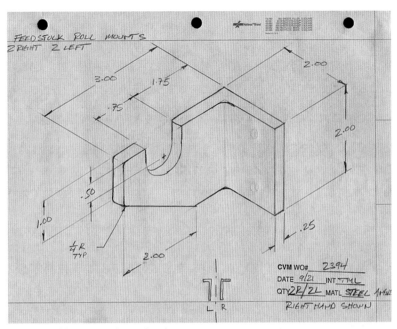

Figure 2–10 Isometric projections.

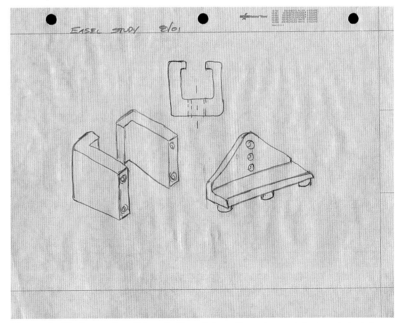

Figure 2–11 Example of the clarity of shop floor isometric hand sketches.

Take time and learn to functionally sketch in the isometric projection (Figures 2-10 and 2-11).

Figure 2–12 An isometric hand sketch.

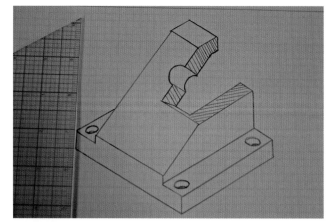

Figure 2–13 Another isometric hand sketch.

Isometric sketches save countless hours of drawing time —many times you need only one view instead of the standard two or three to fully communicate a part (Figures 2-12 and 2-13).

I have a little trick I use to help me with my isometric drawing. I made a special ruler which, combined with normal graph paper or even plain lined paper, can make everybody's isometric drawings look pretty good. You can buy ruled isometric paper through drafting suppliers, fine art stores, and many office supply companies. The problem I have found with the pre-printed isometric paper is this: when you photocopy or scan it, the grid lines get darker and wash out your sketch object lines, making your nice isometric sketch a jumble of spidery lines.

With some common sense and a few guidelines, you can easily make decent looking isometric sketches right on the shop floor. You will need a plastic ruler that's clear so you can see through it (Figure 2-14). It should have lines on it parallel to the long axis of the ruler. These are used to set distances off parallel to existing lines. They also work great for spacing hatch lines when you draw a section view. You can see the Iso-ruler I made is cut off at the same angle as the normal isometric axis, approximately 35 degrees in this case. The see-through plastic and lines are important for setting off parallel distances (Figure 2-15).

If you want really nice circles, you will need an isometric circle template. I skip the template out in the shop and hand draw these. Always remember that the goal is to get it done as quickly as possible with good quality. It's always better to have a decent quick sketch than a perfect sketch that took way too long to make.

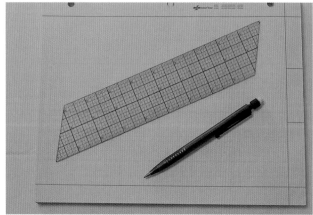

Figure 2–14 See through the ISO-ruler for cross hatching and setting off parallel lines.

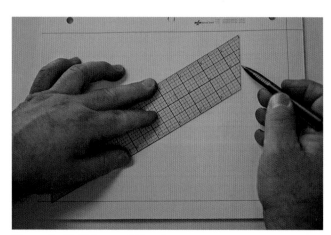

Figure 2–15 Using an ISO-ruler.

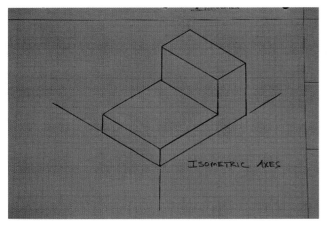

Figure 2–16 The three isometric axes.

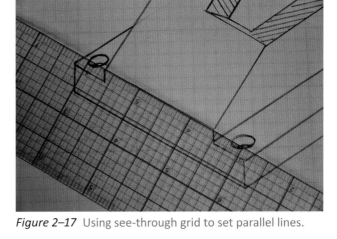

Figure 2–17 Using see-through grid to set parallel lines.

The three axes seen in Figure 2-16 are the isometric axes. When drawing to scale, these are the only directions that measurements can be made along. With the Iso-ruler, you can align the angled edge with the vertical gridlines of normal graph paper or even the edge of the pad. Flip the angled edge over to get an alignment guide for the opposite direction. I like engineering pads that have the lithographed grid on them for anything I will run through the copier. The lines are a faint green and disappear in the copier. These pads are made by National and distributed by the Avery office supply company.

With the parallel lines on the ruler, you can set off distances quickly by sighting through the clear ruler (Figure 2-17). This is a great simple method for drawing quick, concise shop sketches. Just remember: it's better to have some kind of sketch even, if it's not technically perfect. You just can't do this stuff from pure memory.

When quick sketching, try to get the scale and proportions to look right. They don't have to be deadly accurate; they just need to look about right. If a part is long and skinny, try to sketch it that way.

Learn and use the basic rules of drafting. Think you know what they are? Open both a drafting book and your mind. These rules and conventions have been refined over hundreds of years. You can bend them a little, but don't break them into a million pieces.

Some of the major abuses I see in modern electronic drafting are incorrect line weights, omitting hidden lines in views, missing dimensions, incorrect third angle projections, lazy details on hole bottoms, and thread misrepresentations. The list goes on and on. These abuses result from ignorance of the conventions of drafting and plain old laziness. Just because you can draw something with an expensive electronic drawing program does not mean you are doing it according to established conventions. As the saying goes, garbage in, garbage out.

Avoid the temptation to use an electronic drafting or modeling program right out of the gate on a new design. In my opinion, CAD is often a hindrance in the earliest stages of a design—or worse, a waste of time. CAD promotes a microscopic view with way too much detail and precision in the opening stages of concept development. Coupled with the possibilities of countless revisions and iterations, this weakness can make CAD a serious time liability. If you can't get the concept down with paper and pencil, you have no business behind the mouse of a computer.

Keep your original hand sketches in a binder. Having these handy has saved me many times. Always release a copy to the shop and retain the original information as backup.

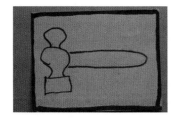

Figure 2–18(a)
Press Fit.

Figure 2–18(b)
Deadly Feature.

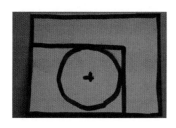

Figure 2–18(c)
Radius Too Small.

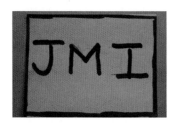

Figure 2–18(d)
Just Make It.

Figure 2–18(e)
Day Shift Feature.

Figure 2–18(f)
Night Shift
Feature.

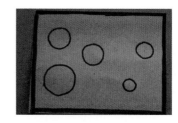

Figure 2–18(g)
Extra Holes OK.

Figure 2–18(h)
Regardless of
Feature Cost.

Figure 2–18(i)
Close.

Figure 2–18(j)
Really Close.

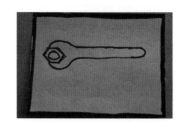

Figure 2–18(k)
Tight.

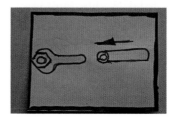

Figure 2–18(l)
Really Tight.

As a subject, geometric tolerancing, otherwise known as ANSI Y14.5, is a whole can of worms on its own. Entire books devoted to the discussion and endless dead dog beatings are associated with this subject. I agree with the overall idea and philosophy, but the execution has been less than perfect. Hundreds of drawings cross my desk every year and I see every misuse of this system in the book. Interpretation and confusion caused by a system that was designed to help eliminate these same issues has wasted countless millions of man hours. It literally has turned into the Swiss army knife of tolerancing systems. Use it carefully and with deliberate thought and purpose. It's kind of like a loaded gun in a crowded elevator; you really want to be careful where you point that thing.

Figure 2-18 shows some of my own personalized favorites. I call it ANSI WISEGUY 14.55.

Hey, if we can't poke a little fun at ourselves, what's the point? I would like to have rubber stamps made and mark up the drawings that misuse the symbols and send them back to where they came from.

You can convert old sketches into new sketches. Many parts are boringly similar. Use this similarity to your advantage when sketching. I have a couple of hand sketches without the dimensions—I copy them over and over to save sketching time (see Figures 2-19 and 2-20). Just fill in the blanks and, whammo!, a new drawing of the same old boring part.

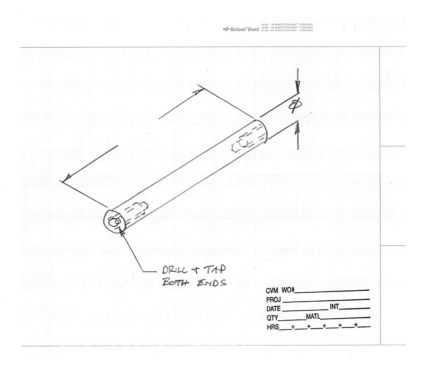

Figure 2–19 Converting an old sketch.

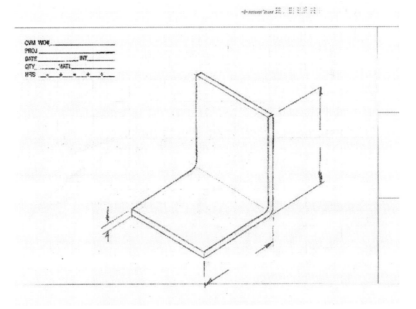

Figure 2–20 Reusing old sketches can save time.

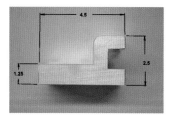

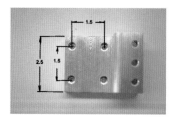

Figure 2–21 A digital picture of a part.

Figure 2–22 Add dimensions to a digital picture of a part.

You can also print out a digital picture of a part, then dimension the photo to create a decent drawing (Figures 2-21 and 2-22). Sometimes you can lay a part directly on the copier or scanner to get a decent print.

Laser printers are so good now that you can print to scale and measure the lines with calipers to double check yourself or find a missing dimension (Figure 2-23). If the designer used a computer, there is a better than 50/50 chance the drawing is correctly drawn to dimension.

You can use this to your advantage to determine a missing dimension even if the paper drawing has not been plotted to any particular scale. If you measure a dimensioned distance, you can calculate the scale of the plot and derive missing dimension pretty accurately. Obviously it's better to have the real numbers, but sometimes you have to get a part out regardless of any missing information.

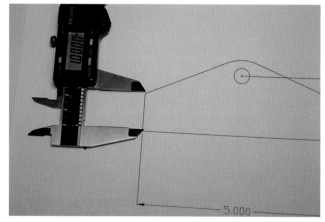

Figure 2–23 Using calipers to find missing dimension from an accurate laser plot.

To save layout time, use light contact cement to bond a full-scale laser drawing directly on your material (Figure 2-24). This is a great prototyping trick. Put tiny circles at the centers of any holes you want. A .015-diameter center hole printed out makes a great center punch target.

You can make a larger 1:1 full-scale drawing with an 8.5 × 11 printer by making a mosaic (Figure 2-25). You can even have little alignment marks to make the mosaic assembly easier. A long, skinny full-scale drawing is a good example of something you can use this trick on. Figure 2-25 shows an assembly printed out 1:1 that would not normally fit on 8.5 × 11 paper.

Figure 2–24 Bonding a full-scale laser drawing.

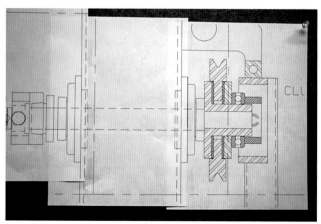

Figure 2–25 A mosaic drawing made up from several smaller full-scale drawings.

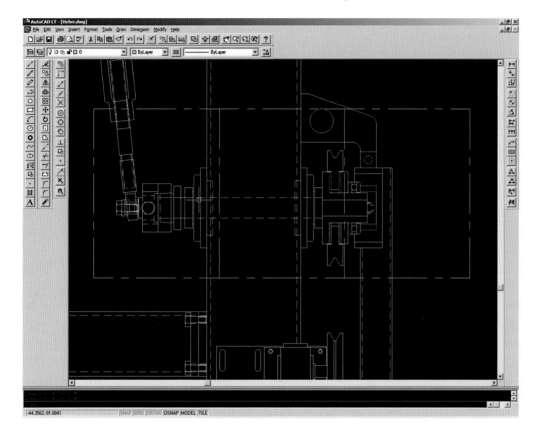

Figure 2–26 Tiling together a large drawing.

Figure 2-26 shows the approximately 8.5 × 11 print sizes overlaid on the drawing. There is really no limit to how large a part or assembly you can tile together using this method. For some jobs having a 1:1 scale, drawing is pretty handy.

Date and initial all your hand sketches. It improves communication and documentation, and everyone will know who is making those great sketches.

A quick five-minute sketch before a rush job has saved many a metalworker from painting themselves into a corner (Figures 2-27 and 2-28).

Use whatever is handy to communicate in a written form. I call this method Table-Cad. I've done some of my best design work right on the table. Just don't spill the acetone.

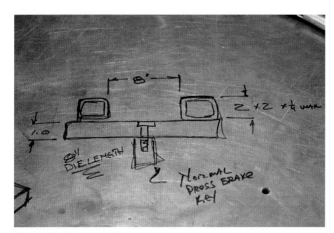

Figure 2–27 A quick sketch.

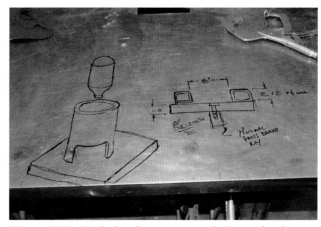

Figure 2–28 Quick sketches save time during rush jobs.

2.3 Minimizing Screw-Ups

Some of the best learning and a good portion of skill development come from making mistakes. Why is it when the job goes smoothly you can't remember how you did it? But when you struggle or make a huge mistake, the memory is much more imbedded and clear. Treat your mistakes as learning tools. It is even more important to know what not to do than to think you know.

There are a million ways to screw up in this business. Learn from other people's mistakes as well as your own. Planning and anticipation of potential problems are hallmarks of people who have made enough mistakes to know better. Think the job all the way through. A few minutes making a plan will pay you back tenfold. Change your plan if it looks bad. Plans should be flexible and dynamic to account for unexpected events.

Admit your mistakes quickly and move on. Dwelling and denial waste time and valuable energy. The difference among mistakes is in how much times goes by before they are discovered. The trick is to catch the mistake before much time has passed. Check frequently and consistently.

Metalworkers often make mistakes on simple jobs but have no mistakes on complicated demanding jobs. Attention is the difference. Pay attention on the easy ones.

When in doubt opt for maximum material condition. Doing so may give you the option of going back and removing more if needed.

Checking your work often with attention can minimize the time a mistake can exist.

Watch out for the right hand left hand or mirror scenario. Many times to save drawing time a mirror image is called out in the notes or work order. Don't fall for this one.

The classic dunce mistake is to miss the quantity. It's the pits when you make one part and three were called out.

When in serious doubt or confusion, go back to something you absolutely know is correct either by direct measurement or observation. Work your way forward from this point. It's dangerous to make any assumptions when in doubt.

When tracking down errors, make no assumptions. Verify everything when doing a forensic investigation. That means you take and read the measurements yourself. Don't just assume that the other guy did his part correctly. Verify it.

When you can't find a direct cause, look at cumulative errors and effects. Small amounts here and there can add up and throw something significantly off farther down the line.

There are some types of work where errors can be catastrophic (Figure 2-29)—moving your own machinery, changing your electrical wiring, perhaps working on a client's rare collector car. As you're developing your skills, watch how others handle these kinds of jobs—learn their methods. If you find yourself responsible for this type of work, slow down and think the job all the way through.

Figure 2–29 Some days are terrific days. This is not one of them.

2.4 Accuracy

Some ways of becoming an accurate metalworker are related to the habits developed in the quest to minimize mistakes. Attention to small details along with deliberate and frequent checking help.

Real accuracy comes from your work habits and skills, not the particular measuring tools.

Don't let the dimensional accuracy get out of control before you take action. Small, measured corrections are always better than gross corrective applications.

You cannot expect to last long if high accuracy is your only goal. Tolerances are stated to allow for material variations and manufacturing efficiency. Learn to use them. You cannot justify producing half the work at twice the needed quality and tolerance requirements just because you can while your work mate churns out lower quality parts that nevertheless meet specification and are within tolerances. In other words, don't build a bridge with a micrometer when a ruler will suffice.

2.5 Speed

One trait that contributes to speed is decisive decision making. Sometimes it's more important to make a decision than to try to make the perfect decision.

If you realize your decision was wrong, make a new one and proceed again. Hand wringing and whining are a brain drain. Allow for mistakes and variables in your plan. Don't find yourself in enemy territory and out of bullets.

Take the long view. Try to anticipate things that will be needed and when they are needed. Don't let the lack of a simple part or tool stall your entire job.

Identify critical aspects of the job early on and have a contingency plan if something changes or doesn't go as planned.

Lead from the front. Demonstrate by confident example. These are the hallmarks of a great teacher and an even better foreman. A leader is also the first person to get shot in a charge out of the trenches. It takes a certain kind of personality trait that some people may lack.

Keep your momentum up. It's easier to keep the job moving with small consistent energy inputs. If it stalls, a much larger input is necessary to get it moving again. Inertia does work.

So now your brains hurt and you want to look at more pretty pictures of interesting tips and tricks, fine with me. There will be a test on this chapter Monday morning........

Somebody once told me a funny story about a shop that was on the waterfront somewhere. The shop had been there for quite some time. The machine shop bay doors opened onto a narrow estuary separated by a wooden dock that ran along the side of the building. One day there was an unusually low minus tide. To the shock and horror of the foreman, littering the exposed mud of the estuary were hundreds of scrap parts. Most of the stuff was piled about the distance you could heave something from just inside the bay door. Apparently when something got screwed up it got tossed into the water hopefully never to be seen again. Whoops!

2.6 Shop Math

Here are a few tricks that can make you look like a genius with half a brain or at least half a head (Figure 2-30).

This is another area where a little practice goes a long way. I've found over the years that many metalworkers hated math in school. To me this seems weird because of how much math there is in the metalworking trade. If you want to be a top dog with your own bed and bowl, you will need some math in your toolkit. This section is not meant to be a complete course in shop math—only a few useful nuggets. I didn't invent any of this stuff; some old guys who had way too much time on their hands thought of it years ago.

I was the same way as many people when I first started out. I didn't like math and avoided it like you might avoid an IRS tax audit. My turnaround came when I found myself in a tight spot—I was tapped to teach a welding class basic shop math! I tried my best to weasel out of the trap but there was no escape; everybody else had already out-weaseled me.

Figure 2–30 Math tricks can help in many ways.

The weekend before the class started, I hit the books pretty hard trying to get a leg up on my students.

What I discovered during that painful period was all I had to do was go slow and stay a little ahead of everybody else. I learned two major lifetime nuggets teaching that first class.

First was that a little practice is all it takes. A strong motivator, like having to chew your leg off to get out of trap, helps a lot also. And the second nugget was the more you practice the better you get.

It sounds lame, but it's very true. Teaching that class several times firmly cemented the knowledge. It also gave me the confidence to go much farther.

I have found that skill in math is one of the major dividing lines between the trades and the sciences. It's more like a stumbling block, or the tar pit that the dinosaur gets stuck in. This one skill has limited more people in the trade than I care to estimate. The worst part is there is no reason for it other than lack of trying.

In reality, there are a small number of concepts and formulas you actually need to memorize. The real trick is to leverage what you have memorized to analyze other problems.

Almost all math problems can be solved in several distinctly different ways. Sheetmetal layout is a great example. You can use trigonometry to figure out the compound angles and true lengths found in many sheetmetal layouts.

Or you can use the graphic methods that are taught to all apprentice sheetmetal workers.

The results are equal in all respects. In fact, the graphic method was taught for the specific reason to avoid mathematics, which was the perceived realm of educated scientists and engineers—some kind of throwback to forbidden knowledge.

The Circle. Let's talk about the circle first. This is one of the most important shapes we will ever encounter in the trades. It pays to know a few of the facts related to the circle.

Every metalworker should know the terminology and properties of a circle. Figure 2-31 shows the basic parts that you should understand without question. These come up so often in our field during the course of everyday problem solving that without the knowledge of these properties we are handicapped.

I have tried to keep some of the more useful properties of the circle within easy reach in my limited cranial memory bank.

A line tangent to any point touching a circle is perpendicular to the center of circle at the point of tangency (Figure 2-32). This fact is useful when doing inspection and reverse engineering.

Two tangent arcs of different radii will have their centers along a common line passing through the point of tangency.

The arc length of any circular arc can be quickly calculated by multiplying the radius by the arc angle by .01745. In formula,

$$r\alpha\,(.01745) = \text{Arc length.}$$

You will see this number again; it's one of the ones worth remembering.

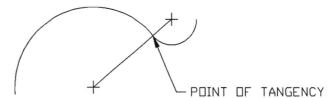

Figure 2–32 The line connecting the centers of the two arcs passes through the point of tangency.

Hopefully we all can remember that the circumference of a circle is the diameter multiplied by the ratio π.

True roundness or circularity cannot be measured with a two-point measuring system. An example we have all seen of this is with poor, center-less ground shafts and pins. These will measure correctly with a two-point measuring tool like a micrometer. But when these pins or shafts are spun in a v-block and compared with an indicator, they show some roundness deviation as runout. The only true way to measure roundness or circularity is by rotation of an accurate spindle or center point.

I have found the formulas relating to measuring chords to be extremely useful (Figure 2-33). I have often used these calculations to find an unknown radius. Many times things in the field or in the shop have a radius that is outside the range of standard radius gages.

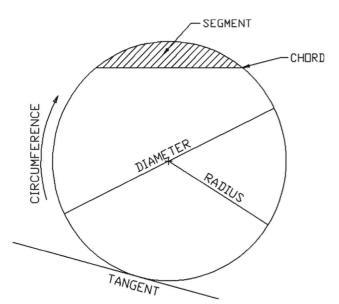

Figure 2–31 Terminology of a circle.

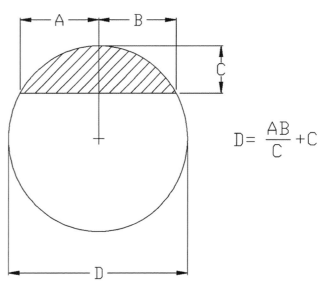

$$D = \frac{AB}{C} + C$$

Figure 2–33 Determining the chord of a circle.

You use a bar or pin of a known length to act as the chord for your measurement. Finding the radius can then be reduced to a simple calculation. The space between the pin and the radius to be measured can be determined with gage pins or by measuring off the top of the pin with calipers (Figure 2-34). Be sure to measure in the center of the pin or bar to get the maximum measurement to the radius. A depth micrometer can be used for this purpose also. The tip diameter of the depth micrometer will affect the depth measurement, but you can add a ball tip to your measuring tool to get around that problem or to make some additional correction calculations for the tip diameter.

Another slightly different formula for chords is,

$$R = \frac{C^2 + 4H^2}{8H}.$$

This one is arranged to give the radius of the arc instead of the diameter (Figure 2-35).

Figure 2–34 Using gage pins and chords to measure an unknown radius.

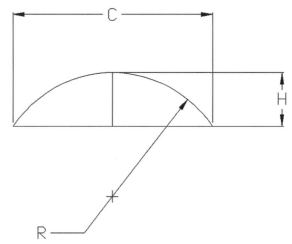

Figure 2–35 The radius of an arc.

The Slurpee Program

At one place I worked, two brothers also worked in the shop at the bench next to mine. They were hired as kind of mid-level mechanics who did a little bit of everything including minor seagull welding and rough machine work.

Now I know it takes all kinds to make the world go around, but these two guys were a piece of work. I never really figured out how they lasted as long as they did. On one level, I was glad that I didn't have to do some of the miserable jobs they were given.

I think between the two of them they might have had two years experience in welding and half that in machine work. They were actually pretty good at some of the real quick and dirty kind of jobs that always seem to be around a metalworking shop.

Fairly often they would ask me questions about how to make different things or how to set up one machine or another to do what they wanted. If you're like me, you go ahead and give them the answers even when you know they should have figured this out themselves by now or remembered it from the last time they had to do it.

One day the older brother came up to me and asked me how to figure the length of a sheetmetal blank needed to make a cylinder of a specific diameter. Now in the sheetmetal world you learn how to do this right after you learn how to use the potty all by yourself. To make matters worse, I had told him at least twice how to do it. I even wrote it on his workbench in sharpie. I guess he spilled the acetone on his bench because he pleaded with me to tell him.

(Continued)

(Continued from page 28)

I guess this was the straw that broke the camel's back, so to speak. I blasted him up one side and down the other. At the end, he actually was begging me to just tell him how to do it so he could get his job done. I caved and told him one more time with the caveat that I would never tell him again. He either had to write it down or ask someone else. He thanked me profusely and off he went.

When I came back from lunch later that day, I found a supersized 7-11 Slurpee waiting in a little puddle of ice cold condensate on my workbench (Figure 2-36). It was the middle of summer, so it was a welcome surprise even though the shop was air conditioned.

After a bit, I saw the older brother and asked him if he was the one who dropped the drink off. He nodded and thanked me again for helping him. I don't know about you, but I find it sure is nice to be acknowledged for doing somebody a favor. It feels downright good. I thanked him for the tasty beverage and thought to myself that if people acted like that more often, it would make it a lot easier to be nice to people. Then, suddenly, I snapped out of that daydream.

I don't quite remember how long it was until the younger brother came up and asked me how to make something he didn't have a clue on how to start and even less of a chance of completing. It must be a very difficult position when you have been assigned something that you know is simple, but you don't know how to even start. And to top it off, you will have to ask somebody who's going to bust your butt before they give you the answer.

About halfway through the blasting, I realized my lips were a bit parched. Offhandedly, I suggested that if he brought me a tasty frozen beverage back from lunch, I would open Pandora's box of knowledge and tell him all the secrets of his latest problem. I was surprised at how quickly he agreed.

I'll give you two guesses what was waiting on my workbench when I came back from lunch. The first guess doesn't count. I had just invented the Slurpee program!!

For a while it was great. I even got to the point where I knew what flavors were available on each day of the week. Out of sheer boredom, I invented complicated mixtures of half grape one quarter lemon lime and the rest whatever red stuff they had, just to try and throw them off. I can tell you these guys had a lot of questions and once they had a system for getting the answers they used it. In actual practice, if you implement a similar system I could suggest a few changes to make it even better. You might try the "Steak Sandwich program" Or the "Fill my gas tank program" instead.

I retired the Slurpee program when I realized I had gained ten pounds and my tongue had an ongoing purple tinge to it.

The moral of the story is this: learn everything you can to be self-sufficient or be humiliated into paying for information.

Figure 2–36 Drawing a circle.

2.7 Mass, Volume, and Area

Another math topic that always seems to crop up is the measurement of weights, volumes, and areas. The typical questions go something like this.

How much does that plate weigh? How much would it weigh if we switched to aluminum instead of steel? How many gallons does that tank hold? Or, perhaps, how much does that tank weigh when it's full?

If you keep a few common facts in mind, you can handle these problems on the fly when they come up. The real trick is to try to remember only a few key items that you can keep in your head and leverage to derive other things. I'm pretty happy if I can figure something out in my head and be within 10% of the correct answer. A few that have seemed to stick in my head and be useful over and over again are:

- There are 231 cubic inches in one gallon.

- Water weighs 8.33 lbs per gallon. Or the shorthand method, "A pint is a pound the world around," which gives the approximation of 8 lbs per gallon.

- Aluminum is about 1/3 the density of steel. Most aluminum has a density pretty close to .10 lb/cu", depending on the alloy. This is easier to remember than the exact number. It's also an easy factor to divide or multiply by. If you want to know the exact numbers, 6061 is .098 per cu in., 2024 is .101 per cu in, and the rest are so close it's almost irrelevant.

- 1 pound per square inch water pressure is a column 2.31 feet tall. This is easy to remember; the numbers are the same as the number of cubic inches in one gallon.

- 1-inch-thick steel plate weighs 40.8 lbs per square foot. I round off the decimal and just use 40 lbs per square foot. This gives an answer accurate to 2%.

- The density of steel is approximately .283 lb/cu". You can round this off to .3 and be within 6%.

- The United States five-cent coin (the nickel) weighs almost exactly five grams. It is so close, it's scary.

2.8 Angles and Shop Trigonometry

I've mentioned before in this book that angle work can be some of the most difficult work facing a shop worker. Laying out and measuring angles on the vast number of projects that you might encounter requires a good grounding in trigonometry and geometry. You can get the necessary background from a large number of other books; our focus here is how these topics relate to your work in your shop actually building things.

This section includes a few of my personal favorites that I use over and over again.

SOHCAHTOA. The way I learned basic trigonometry concepts was through the story of the old Indian chief Sohcahtoa. If you remember him, you can easily solve any right triangle problem you might bump into during the course of your work. Here is how it breaks down if you have not seen this before.

SOH = Sine θ = Opposite/Hypotenuse
CAH = Cosine θ = Adjacent/Hypotenuse
TOA = Tangent θ = Opposite/Adjacent

That's it. Memorize this and you will have right triangles nailed (see Figure 2-37). Another thing some people haven't learned is to calculate direct angles by using their calculator shift keys and the appropriate trigonometric function for their problem, which on most pocket calculators raises the function to the power of negative 1. This one extra step will display the actual angle instead of the trigonometric function of that angle. The key will generally be—TAN^{-1}, SIN^{-1}, or COS^{-1} (see Figure 2-38).

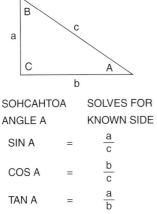

SOHCAHTOA   SOLVES FOR
ANGLE A   KNOWN SIDE

SIN A   = $\dfrac{a}{c}$

COS A   = $\dfrac{b}{c}$

TAN A   = $\dfrac{a}{b}$

Figure 2–37 SOHCAHTOA.

Figure 2–38 The SIN, COS, and TAN buttons of most calculators also let you calculate SIN^{-1}, COS^{-1}, and TAN^{-1}.

Square Root of 2. Another number I have found to be extremely useful is 1.414, the square root of 2. The ratio 1.414:1 is valuable when finding angles and lengths related to 45°, which pops up in the shop on a daily basis. Remembering 1.414 is very useful for figuring chamfers, spot drill diameters, and the cut lengths of gussets, diagonals, and braces.

Angular Divisions. When I used to shoot at the target range, I learned how small some circular divisions are. The adjustments on telescopic sights for rifles normally move in 1/2-minute divisions with some going even smaller. At 100-yards distance (the length of a football field), 1° of angle is 60 inches (Figure 2-39). A good shot with a decent rifle can shoot under 1 minute of angle (1/60 of a degree or 1 inch at 100 yards). One degree has 60 minutes of angle and each minute has 60 seconds of angle.

Our 1 inch corresponds to 1 minute of angle at 100 yards, so 1/60 of one inch is 1 second of angle at 100 yards distance. In turn, 1/64 of an inch is approximately 1 second of angle at 100 yards. I have 64ths divisions on the pocket scale I carry; that's a pretty small angle. The Moore Precision Tool company makes an 8-inch rotary table accurate to 1/100 of an arc second—yikes! That about .0002 inch at 100 yards!

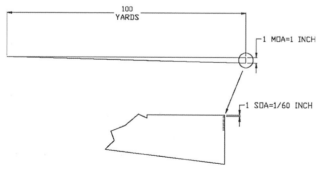

Figure 2–39 Angular divisions.

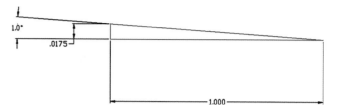

Figure 2-40 Graphic illustration of the relationship of the sine of 1 degree.

Sine of One Degree. This little trick is handy for calculating angles mentally and rises when you don't have a calculator handy. Its basis is the sine of 1 degree, specifically the sine of 1 degree at 1 inch length is .0175 (Figure 2-40).

This ratio of .0175 per inch scales quite well. If you increase the angle to 2 degrees, you can double the .0175 to get .035. All you need to remember to apply this handy trick is .0175 is the rise of a 1-degree angle at 1 inch in length. If you double the length, then double the rise. If you increase the angle by five, then increase the rise by five.

It scales well up to quite a few degrees of angle. I generally use it only for estimating small angles under 10 degrees.

What is the rise of an angle of 2 degrees at 12 inches? Using this method we would double the sine of 1 degree (.0175) to get .035 and then multiply .035 by 12 to get .420. If you ask the old Indian chief Sohcahtoa and apply some simple trigonometry, you get .419, which agrees closely. Try it yourself with a few angles to assure yourself of the scaling. I have found this to be a very useful technique. A typical question you might want to answer using this trick is: what is +/− 1 degree at a 10-inch radius? .0175 × 10 = +/− .175

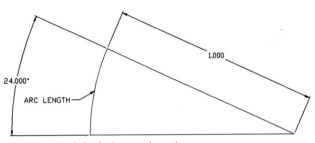

You can expand on this technique to calculate arc lengths:

$$(r) \times (\theta) \times .01745 = \text{arc length}$$

For small angles, the rise and the arc length are almost the same. This technique works best for calculating longer arc lengths where the angle is much larger. For example, in Figure 2-41, the arc length is

$$(1.0)(24)(.01745) = .419$$

Figure 2–41 Calculating arc lengths.

2.9 The Metric System

Can we talk about the metric system? I know it is supposed to be better and easier to use. My gripe is being caught in a change that will realistically take the United States 40 years or longer. Some industries are reluctant to switch because of the huge investments in infrastructure dedicated to imperial measurements. How can we be expected to change when the basic infrastructure is so deeply reluctant and lacks any incentive to comply? Until I can get materials, drawings, and supplies readily in metric sizes, I will keep one leg firmly planted in the English world. If we are forced into compliance, it will only cost us—with no return on the massive investment to change. The folks working outside of the required compliance net will reap the benefits of our added overhead; they will make parts and pieces as good, or better, than we do at lower cost.

Problems come up in engineering and metalworking when we switch between different units. The effects manifest themselves in details that get messed up anyway like tolerances and surface finishes. Alarm bells go off when you see really tight tolerances on otherwise mundane metric drawing. There is a huge difference between .005 inch and .005 mm. I just want good drawings.

In reality, the metric system holds no particular advantage to the machine tool industry. Yes, in some sciences and true engineering, the metric system makes sense because they are manipulating numbers to obtain a solution and unit conversions are simplified. In the machine tool industry, we merely read whatever numbers we are given and make the parts accordingly.

Think about this a little. The machines read the numbers they are fed and have little meaning to the machine programmer or operator. Who looks at 50,000 lines of numbers with any comprehension? If we use decimals instead of fractions, we are using a base ten system that looks suspiciously like the metric system. In fact, you cannot tell the difference. Is the gripe with Imperial really only related to fractions? No, the real problem is units. The basic stumbling block has always been and always will be units. Hey, we put guys on the moon with slide rules and the English system—sounds pretty good to me.

Are metric measurements easier to record? No. They are just numbers we read from our tools and compare to the drawing. If the numbers on a drawing were replaced by letters or symbols, could you still make the part? Sure, as long as all the letters you needed were on the drawing and your tools were calibrated in those letters. So we're back at the need for *good drawings*, not a particular system of measurement. The metric system cannot help with the problem of bad drawings.

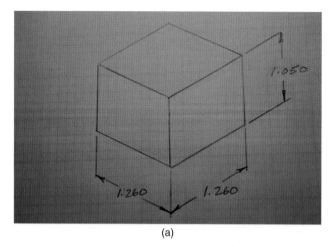

(a)

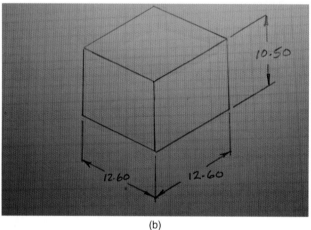

(b)

Figure 2–42 Which drawing is metric?

In Figure 2-42, guess which drawing is metric. Everything is meaningless—including the metric system—unless we specify the units.

To illustrate how confused everyone is, I saw a kid on TV asked how big a particular dinosaur was. His reply was "Fifty meters long, and 130 feet high." He probably memorized these facts from two different dinosaur books and went ahead like a good future engineer and mixed the units. In itself that is not a problem; at least the kid gave us the units he was using, which is better than most of the drawings that cross my desk with the dimensioning mode set to "free for all."

A friend who is an aspiring metalworker told me that when he needs to take really "accurate" measurements, he switches to millimeters.

I think because the lines are closer together on his plastic ruler he stole from the kindergarten kid down the street, it just feels more accurate to him. I don't have the heart to tell him how those pesky little 100ths are even harder to read. I think he also has a pair of those scissors they give you in fifth grade with the rounded ends to go with his thermoplastic ruler.

One of my favorite metric stories is about a job a friend did in his shop for one of the big government labs. They gave him a nice detailed metric drawing to manufacture a part. All the dimensions were on the drawing so no questions ever came up. When the part was delivered, the "scientist" asked why the part was so small. Well it turns out he was using centimeters to draw the part and it was manufactured using millimeters. Whoops, Houston we have a problem, the satellite just burned up in orbit….

I don't really love or hate the metric system. I just don't like having the metric system rammed down my throat by somebody telling me it's easier to use or we won't do business with you unless you are "metrified." I'm what you could call "prochoice" when it comes to the metric system.

So the next time somebody tells you how much easier the metric system is, quickly ask them how tall they are or what their body temperature is. If they reply in feet and inches or degrees Fahrenheit, you can then send them away to calculate their height in some metric units.

At the same time, ask why, if the metric system is so easy, they didn't just use it. Or if they reply with metric units and ask you what your height or temperature is, reply in some unexpected metric units like nanometers or degrees Kelvin and see what kind of reaction you get. Hey they're part of the metric system—just convert it. So easy.

2.10 Computers and the Metalworker

Every day computers gain more foothold throughout every industry worldwide. No doubt computers have changed our planet forever. We can argue whether the change is for the good. Many metalworkers shy away from computers and machinery controlled by them. Why is that? To modern metalworkers, the computer is just another tool that allows them to do a different class of work or the same class in a different way.

In most cases, computers are an age barrier. There seems to be a dividing line in ages between computer acceptance and disdain or outright rejection. (This is not concrete; there are many crossovers.) As a metalworker, you need to learn a minimum of three skills with the computer or go the way of the ice cutters (see box). Start by thinking of it like a tool in your toolbox.

Tale of the Icecutter

Before the invention of electric refrigeration, there were professional ice cutters. In the winter, they would go to the frozen lakes and saw blocks of ice for use during the rest of the year. The ice was stored in insulated barns and caves. At the time, the craft had evolved into a mature and efficient industry, with special lifting and transporting equipment to cut, lift, and move the blocks efficiently. During the rest of the year, the ice cutters sold blocks of ice to households and businesses to keep food and products cool. The term *icebox* is related to this time period. They readily accepted technology improvements that improved existing operations—up to a point.

Then along comes someone who discovers that expanding gases and a compressor can produce cold and even freezing temperatures without ice. The ice cutting industry rejected the technology, partly because it directly threatened their livelihood. The technology was new, radically different.

Computers are like refrigerators in this story. No longer new, they are still mysterious; they often seem complicated and can be frustrating. They directly threaten many people's livelihoods. And they're here to stay. As with any new skill, the first part is the most painful. I can say with a straight face that learning how to use a computer is a lot less painful than some of the other skills I have learned over the years.

Learn to draft on the computer. If you decide you want to learn only one thing, drafting would be it. It doesn't matter which exact program you learn, only that you learn one well enough to function with it and make a good drawing. Once you have learned one, transferring to another program is a thousand times easier. You already have a major head start in electronic drafting because you know what a good drawing looks like and the names of the different elements found in an engineering drawing. And by deciding to learn electronic drafting, you are already way ahead of the kid whose last job was making sandwiches.

Keep in mind the microscopic effect that computer design fosters. Computers cause you to focus on too much detail early in the design process. In the early stages, a pencil and paper are the right tools. If you are scratching your head with a pencil on how to do something, you should think twice about jumping on the computer.

Learn to make, take, send, receive, and edit digital pictures and files. Things have sure changed since I started in this work. People didn't have powerful miniaturized computers strapped to their hips like they do now. Nowadays, if you don't have the latest, most powerful phone, or some pad or pod, you can look up your picture in the dictionary under "Metalworkersaurus." If you don't have one of these fantastic, modern tools, then you are missing out on a much larger world. Every day you delay brings you that much closer to your "sell by" date.

A picture is worth a thousand words. I think everybody has heard this one before. So if you have a picture *and* a thousand words, what's that worth? Even the ability to ship drawings around the world in one day pales in comparison to the speed of electrons inside copper wires or optical cables. Today, one day (or less) is the difference between getting the job done and missing out completely.

Do yourself a huge favor—get a digital camera and let the learning begin. You will need to learn not only how to take the pictures, but also how to do minor editing and re-sizing. This skills will only make your experience that much better. You can share and exchange vast amounts of information around the globe using the Internet and your new camera. You might meet a Danish sheetmetal worker whom you introduce to your New Zealander panel beater friend. Or you might develop a friendship with another inquisitive tradesperson looking for the same answers as you from eight time zones away. The connections are limitless.

A handful of years ago you had a very slim chance indeed of meeting any of these people, let alone learning something from them or interacting on any level.

I have personally traded parts for a lathe for a pail of pickled herring with a nice person living on an island in a fjord in Norway through the power of the Internet. Thanks Ole!

Learn to search for information and resources on the Internet. Every day I am amazed at the information that is available to us. I am also appalled at the amount of garbage that circulates with it. Fortunately, we can still filter better than any computer.

- Weigh the efficiency a computer can add to a task. They don't automatically add value to every job.

- It's often hard to see a big picture (or the bigger picture) on the monitor. At some point, you have to make something gin the shop. A quick mockup can save you many burned-out eyes in the office.

- Many welders and machinists are so fast they can create something in the shop for testing before it can be drawn up on the computer. Don't underestimate these guys and their abilities. They can test three ideas before another can detail one on the computer. There is no substitute for a physical prototype; it has a presence like no drawing ever has.

- If you're not diligent in how you use computers, you can find yourself stuck in the details when all you wanted was to check a dimension. Don't miss the forest!

- Be realistic about the information the computer gives you. It may calculate dozens of decimal places but that does not assure accuracy if your inputs are wrong.

2.11 Dumb and Dumber

This is a little steam vent section where we can all have a little laugh at some of my pet peeves. I'm sure we all have special ones of our own, but this is my book and I get to rant a little and possibly plant an idea of two in a few folk's heads. Don't worry—this is a short section with some funny stuff in it.

Design and engineering can change our lives for the better or be criminally lame. Advertising and marketing blur the line. As consumers, we are often excluded when products are developed. It seems the legal department takes on the bulk of new product design. How many interesting products never get to market because of worry about lawsuits? Profit by litigation weakens our country, once known for innovation and risk.

Product design is such a subtle art. One or two bad decisions can make an otherwise good product bad or, worse, annoying. Well-designed products and tools are a pleasure to use and are immediately obvious in superiority. This goes double for superb personal service. The smooth operation of a fine mechanical device and the silent mind-reading waiter in your favorite restaurant are the results of deliberate attention to detail and practice of your particular trade.

The pinnacle of Elmer Fudd engineering is realized when product designers spend their time wrapping products in a bright finish with fancy, organic, industrial designs. Unnecessary complexity is often included just to claim awards for most features.

Just make the thing work like it's supposed to! A tool should look, act, and feel like a tool. Can we just agree on that? Maybe I'm not the best person to judge aesthetic design, but I sure as heck can judge good function. Remember: form follows function, not the other way around.

Here are some of my peeves. As king, I'd banish them from the face of the earth.

Figure 2–43 The joy of parking lots!

Parking lots. All the rules of the road go out the window without any sidewalks (Figure 2-43). Pedestrians need armored personnel carriers to make it from the parking lot free-for-all to the stores. Most parking lots seem more like holding pens for cattle. Parking lot designers should be condemned to wander blindfolded and naked through their creations pushing a shopping cart missing two diagonally-opposed wheels! All while frustrated parkers encourage them with cattle prods.

How about safe walking zones where cars can't get you? A clever designer could even make these lanes automatically return the carts to the front of the store. Better yet, have your own personal cart that folds up and attached to your car —yours to customize and equip to fit your lifestyle. And you don't have to return it!

Metal paint can lids. Enough already—it's time for a new design! You shouldn't need two tools to open and close a consumer product. Crusty lids and obliterated labels—be gone! Happily, change is finally coming!

Self-checkout lines. This idea almost offends me. I have a choice when and where I exert my economic power. If more people exercised their options when they make a purchase, we would see rapid improvement in this area. When I walk into a business I am three-quarters of the way toward spending my money. Car salesmen are acutely aware of this fact. Why do you think they pounce on you like a cat on a small, blind, three-legged scurrying mammal as soon as you set foot on the car lot?

As I make my purchase selections, I often must make small compromises as I go. Now, I am ready to spend hard-earned economic horsepower residing in my wallet and they want me to check myself out? Who the heck thought of this? I can just imagine a bunch of MBAs sitting around high-fiving each other when somebody suggested this one. Most megastores can barely organize their shelves so you can find things. What makes you think they have fully considered the self-checkout system? We have to draw the line somewhere. If we let the corporate collective do it, I guarantee it will mean more work for us. If they really wanted to make things convenient, you could fill up your cart and just wheel right out the door without stopping, your account debited with 100% accuracy and zero waiting.

What's next, self-service parachute packing, or maybe do-it-yourself laser eye surgery? In our modern world, there is still room and need for specialists and expertise in all industries. One morning we might all wake up and discover to our horror that we don't know how to do anything except consume like beached whales or sell each other cheap junk made in another country that does remember how to make things. I'm sure we can all relate to what a pleasure it is to see a job well done no matter how simple it might seem to an outsider.

The scissor jack on my wife's minivan. I know these are designed to be manufactured as cheaply as possible and not intended for everyday use. Still, they are borderline next-to-being-useless, scrap-barrel material. The handle has no ratchet and can be swung only through a pesky little arc before you have to flop it over to the opposite side. The screw is as free running as a snow plow in a sand dune. To get the jack remotely near the frame of the car takes hundreds of flip flops of the pathetic handle. Meanwhile, you're crouching like a rodeo clown at a bull riding competition trying not to get killed on the roadside. Then you think you're smart by leaving it extended when you put it away so the next time you only need to crank it a few turns. Well, the smarty-pants designers made the spot where the jack is stored fit the fully-collapsed jack like a latex glove that's two sizes too small. I guess I will be sure to renew my AAA membership, forever!

How many times have you seen these jacks tossed into the trunk because they are too much trouble to put back where they belong (see Figure 2-44)? Why can't these things be built into the car chassis? Push a button, or insert a handle into a hole in the body and off you go.

Figure 2–44 A better use of a scissor jack.

Music CD packages. The little tab says "pull here." It should say "pull here to frustrate consumer with immediate seal failure and mandatory sticky goo application." I fail to understand why these packages have to be hermetically sealed. Is it just theft, or just a test to see if you have what it takes to outthink the packaging?

Some folks might say it takes a little practice. Well, I don't have to practice opening the CD player to put the disc in to play it. Why should I have to "practice" opening the crummy ill-conceived packaging?

The lame cart that your welding machine comes with. I don't think designers ever lifted a 330-cubic-foot argon cylinder up the 12 inches to get it over the lip of the cart and into the "convenient" bottle retention area. It's like they built the machine and then said, "Holy cow! We forgot to put a place to hold the gas cylinder. Quick! Get the duct tape and glue gun and add one on before we put it in the box." How about one that lowers to the floor so you can roll the bottle right into place?

People who insist on calling extruded bar stock "billet". Look up billet in your metal supplier's handbook. Anybody who uses this term to describe anything other that a real billet should have a big red letter "D" stamped on their forehead for "dunce." This term is inaccurate and nearly offensive to anybody who knows better. How about machined from slab or bloom? The next time somebody who knows better misuses this term in your presence, just slap them! I'll mail you a quarter.

"Aircraft grade aluminum". What are people trying to say? Airplanes have almost the full spectrum of aluminum alloys in them. Get a little more specific, folks. The trade is full of exact, precise, scientific, and engineering terms. This is purely a marketing ploy.

Electrical plug strips. Why can't designers actually measure the bulbous transformers mated on the ends of all modern electronics? It's bad enough they don't trust us with high voltage 110 anymore. The pitch between the electrical receptacles should be such that transformers can reside in peace with a modest side yard with their neighbor transformer. Or better yet, build the transformers internally with the plug strip and use a small standardized connector for the peripheral equipment.

Modern texts that still have trig tables. Let's pull ourselves by the bookstraps into the electronic age. In the time of five-dollar pocket calculators, trig tables are just fluffy filler material.

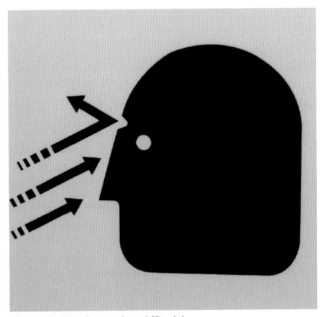

Figure 2–45 It's not that difficult!

2.12 Want to Make a Million Dollars? Some Things That Really Need to be Invented

Here are some ideas out of my personal "Big Book of Million Dollar Ideas." Go for it if you like any of them. I don't have enough time left to work on a tenth of the things that need to be invented to make my life easier.

- Real time coolant refractometers. These would be plumbed into a machine's coolant system and give alerts when coolant concentrations are not within settable parameters. Put a knob on the front of the machine. Heck, the machine can control the entire thing for all I care.

- Built-in oil skimmers for CNC machinery. They all know about tramp oil contamination. Why don't most machines come with one already installed?

- Cheap easy memory upgrades for CNC machines. Only recently have the machine tool builders addressed this with larger memory capacity as standard equipment. Gee, it took only thirty years to get it done. The prices should be on par with PC memory upgrades. This is a great opportunity for the computer geek crowd.

- Carbide inserts optimized for top performance and chip control on plastics. Ever try running your lathe unattended turning Nylon or UHMW? Didn't think so.

- Angular digital readouts for manual milling machine heads. These would also allow you to monitor the tram condition of the head at a glance.

- Right angle attachments for manual mills with a quill feature. Ever try tapping with the right angle head?

- A cheap direct pressure indicator for manual vises and chucks. This would give the operator some valuable feedback. Or at least help calibrate the apprentices.

- Computer controlled active chatter cancellation. Sympathetic frequencies injected into the spindle under precise computer control used to cancel chatter and squealing. Somewhat like active noise reduction headphones. Mechanical engineering students in need of a graduate study subject please apply here.

- A one-eighth-inch-diameter super-rigidium end mill that can cut four inches deep for all those designers who don't bother to think about how things get made. As soon as this is invented, some numbskull will need it to reach 4.2 deep.

">

Brain Food 41

Some other things I want to try. I always thought it would be interesting to completely switch careers once in a while. Not many people who have established a foothold in one trade get to branch off and try an entirely different career. I think there is much to learn about your own trade by examining the work and skills of other disciplines. I also think that this would be a two-way street, with both parties coming away with some new ideas to think about and push the boundaries out a bit.

One way I thought of doing this was to offer a week of vacation and 40 hours of free labor to learn something about another craft. Just hanging around with open eyes and mind, you would absorb a huge amount of information. Is it worth an investment of 40 hours of your life? I think it would be. Just trade a month's worth of lousy television programming for a new set of skills.

Shop ideas I want to try sometime when I have a minute. I thought a boring head in the tailstock of the lathe might make a good taper offsetting fixture. Or it could be used to align a worn tailstock to bring it on center. I have never tried this, but it has hung around in my head for quite a few years now.

I have never done any metal spinning. The process looks very interesting to me. I saw a picture once of the titanium shell for Jacques Cousteau's underwater saucer being spun red hot in a hydraulic metal-spinning setup.

Modify a few different files to fit in the clamping mechanism of a cordless reciprocating saw. If you found one with variable speed, this might make a nifty little addition to the toolbox.

A few careers I wouldn't mind trying:

Tugboat operator.	Stone carver.	Musical instrument maker.
Wood boat builder.	Surveyor	McMaster Carr order filler.
Astronomer.	Glass blower.	Physicist.
Industrial photographer.	Biologist.	Metal spinner

If anybody out there reading this willing to trade some teaching for free labor, drop me a note and let's see if we can work something out.

Bean Counters Lounge

3.1 Engineers and Metalworkers
3.2 Shop Talk
3.3 Dimensioning
3.4 Other Tips

If you're an engineer or designer and you work in a place that has metal-working facilities under the same roof, you are very lucky. It is both rewarding and fun to see your design come to life right before your eyes, especially if it works. There are some challenges inherent with working with shop personnel they don't teach you in any school. I hope you will find useful ideas in this chapter, whether you work in a small or large shop, or have a one-person operation.

In all my years as a metalworker, I have worked with dozens of engineers and scientists. From the metalworker's perspective, this experience can be very rewarding or pure bamboo-under-the-fingernails torture. I can name on the fingers of two hands the engineers who earned my admiration and full respect. Not to say that the rest were bad—just that the really good ones stand out in comparison. I nominate a special place in engineer Hell for the truly bad ones. This is the Hell where it's always cold and noisy and the only work they get to do is lay out the parking lot sprinkler diagram.

The hallmarks of these successful professionals were a combination of ability, empathy, and respect. I think they understood that respect is something that flows in both directions and is really all most people are looking for in an equitable exchange. If you can earn the respect of the shop people, they will truly bleed red blood for you when the chips are down and you need it.

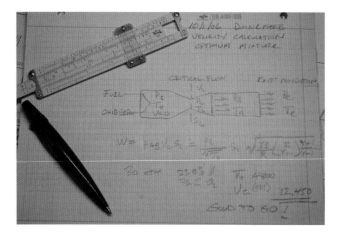

3.1 Engineers and Metalworkers

Engineers and metalworkers seem to come from different sets of molds. Understanding the basic differences goes a long way toward understanding one another. And as with most things, I happen to have an opinion. Engineers and designers are created for the most part in schools with their final luster coming from their first character-building jobs. Metalworkers, on the other hand, have learned a smaller part of what they know in schools, but the bulk of their career knowledge is learned on the job.

The world of engineers is a more open and collegiate environment when compared with the typical shop surroundings. They ask questions of one another in meetings and review each other's work, looking for errors without placing blame. It's a cooperative environment where it's okay to say, "I don't understand" or "I'm not following you."

The world of metalworkers is different. Many metalworkers have a modest-to-medium schooling background. When they make a mistake or do not understand something out in the shop, the reaction is typically different. Someone is usually upset and the individual who made the mistake takes the brunt of the blame squarely on the nose. Or, worse, their workmates relentlessly remind them of the specifics of their ignorance for the rest of their natural lives.

The typical reaction is to hide or minimize all errors. It's simply a matter of survival.

Knowledge and special skills are the measure of a metalworker's self worth. It is quite normal for shop people to keep secret the specifics of their skills and tricks. This becomes the gauge of their value and standing in the shop pecking order. Shop people quantify their performance in physical tons of completed work on the pallet and the bottom line on their paychecks. Kind of like the first tribe to figure out how to make fire. For a while, they were without peer and at the top of the heap. Then some missing link pre-engineer scratched a diagram in the dirt of how to make fire and all the fun was over.

This behavior can be seen throughout history. Engineers and designers write books, take notes, and make drawings. Metalworkers pass their knowledge on to the apprentice in the traditional manner by demonstration and lots of yelling.

If you are lucky to visit another craft person's shop, try to gain insight into the mind of a fellow metalworker. There's much to be learned, even if the shops are cold and smelly. You will find that the more time you spend out in the shop, the more you will appreciate the work that goes on there. Your standing in a large shop, or even a one-man shop, is related to your understanding and empathy for the problems and frustrations of the people that work in them.

3.2 Shop Talk

When working on a new design, talk to the shop people who will be doing the work before you get too far along. They will see things you may never consider. Even if you're preparing a design for yourself, talk to others to get their feedback.

Promptly return any borrowed tools, no matter how small or insignificant. They should all be boomerang brand, and always come back. I cannot overstress this point. The better you are at returning tools, the more likely others will let you borrow again.

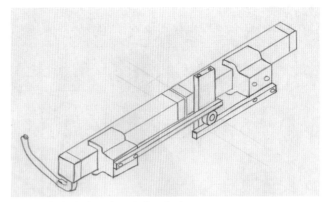

Figure 3–1 A clear hand sketch demonstrates your abilities.

Sometimes it's much easier to tell the shop what you want than to try to over-specify or detail it exactly. Press fits, sliding fits, and threads are a good example.

Impress the shop with a great hand sketch (Figure 3-1).

Do your tolerance analysis and be realistic about your requirements. This area marks one of the great all-time abuses of shop resources. If you don't have time to tolerance with this kind of thought, then leave them off all together or at least ask the shop what can be done realistically.

Allow cleanup cuts on stock sizes if the design will allow it. Your related tolerances should give fabricators a choice. A plate that is nominally .50 thick may have a note, .485 minimum. This allows a cleanup up to.015 if the raw material is rough or comes in a bit undersize. If the drawing calls out careless .500 with a title block tolerance of +/–.005, then your part just tripled in price.

Include stock material variances in your tolerance study. Half-inch thick plate or bar stock is rarely .500 like I see on almost every drawing that crosses my desk. Engineers and designer often have no intent of machining or surfacing. A quick note next to the dimension .50 (STOCK) is a great way to communicate the proper intent. Only ask for maximum speed and effort when you really need it. There is no faster way to wear out your welcome than to abuse this powerful tool.

Respect the shop's time. Friendly yak time can easily get billed to your project. Don't try to help others unless asked. Metalworkers can be territorial and not only bark, but occasionally bite.

Your drawings and instructions are your calling card out in the shop. Make them look good or, better yet, perfect (Figure 3-2). Lousy drawings and bad instructions are a hard reputation to break once you get it. I had one top notch designer who offered me lunch anytime I could find a mistake or omission on any of his work. It was five years before I collected.

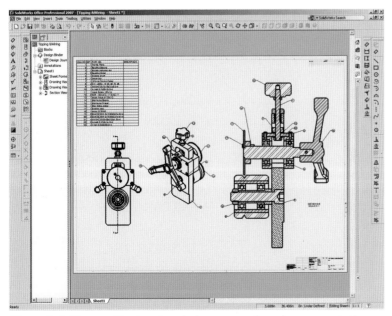

Figure 3–2 Top quality drawings help your work and your reputation.

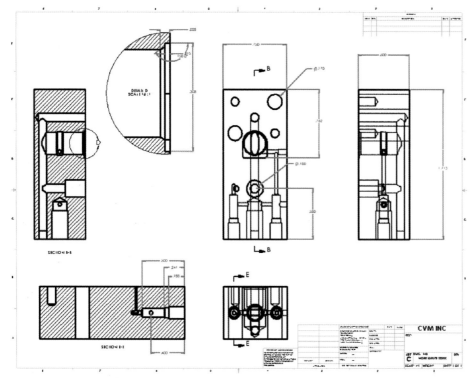

Figure 3–3 Using a breakout detail to clarify complex geometry

Instead of dimensioning to hidden lines or features, try using a section view or breakout detail view (Figure 3-3). This is particularly easy if you are solid modeling.

Add an additional drawing sheet instead of cluttering one drawing page with a million overlapping details.

Include the dimension to the theoretical sharp intersection of any angular relationship (Figures 3-4 and 3-5). For that matter, dimensioning any acute angle or knife edge is a pretty dodgy thing, unless the angular tolerances are such that they include the possible variation of the actual knife edge. Try to avoid these if possible.

Check tolerances extra carefully when converting or using dual-inch and metric dimensioning. There is a big difference between .005 inch and .005 mm.

Include others in critical manufacturing decisions. Having extremely specialized material and process knowledge, they will spot things that are of concern from their viewpoints.

Have a regular presence in someone's shop, if not your own. It's easier to get answers to your questions when you take more than a passing interest You may have had a semester of shop experience, but most shop people have decades under the bridge. Use this to your advantage.

Include the hidden lines in your different drawing views. Turning off hidden lines is a trend I've noticed since the widespread use of solid modeling. You may think it looks clearer but it's more like a game of "Try to guess what I'm drawing wearing a blindfold while standing on one foot."

When important information must be included in note or text form, attach a leader flag to a prominent pertinent object lines in the drawing and reference the note in it.

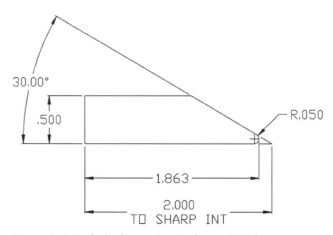

Figure 3–4 Include dimension to theoretical sharp.

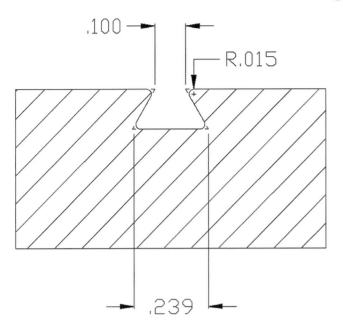

Figure 3–5 Include dimension to theoretical sharp.

Go easy with welding. Massive over-welding is a very common and expensive practice (Figure 3-6). The complications from over-welding are time consuming to correct, let alone the extra time required to do the welding.

How much is too much? One quick rule of thumb is a weld with an effective throat equal to the material thickness is equal in strength to the material. Anything bigger than this is overkill. This is a generalization. The intended usage and design obviously play a part. This is why you went to school all those years—so you could figure this stuff out.

If you make drawing like the fine example in Figure 3-7, don't complain when you don't get what you thought you were asking for. Looks like somebody spilled their spaghetti sauce on it.

Put the o-ring size numbers on the drawing near their respective grooves. Many times the shop has to test or leak-check a part. Having this information readily available on the drawing speeds things up. This information also communicates the grooves usage as a seal surface to the shop.

Use double dovetail o-ring grooves instead or single dovetails if you must use a retentive type groove. This gives the shop more flexibility and options for producing the groove. Single dovetails require fancy footwork on the mill or minor pain in the lathe. Some parts don't lend themselves well to lathes.

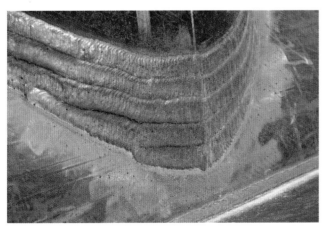

Figure 3–6 Big weld.

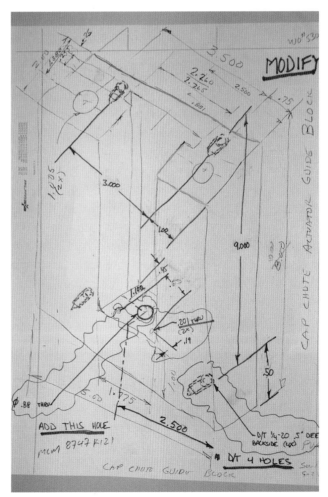

Figure 3–7 This design is hard to read!

Figure 3–8 A different way to round tab ends.

3.3 Dimensioning

Figure 3-8 shows a better way to round the ends of links, tabs, and eyes. A full radius is very sensitive to part width and looks lousy if it's not done perfectly. Two radii are simpler and easier to make look good.

The large radius is equal to the width dimension of the link. The two corner radii are approximately 1/8 the size of the large radius (Figure 3-9). These dimensions are not anything magic, but are offered only as a way to get consistent results with all the different sizes of tabs that come up.

Try ordinate dimensioning for radial patterns (Figure 3-10). This can save setup time and shop calculation errors. X and Y coordinates are more accurate than angular layouts for large diameter patterns. Most of the time the shop converts them back to ordinate. This conversion process introduces yet another chance for error.

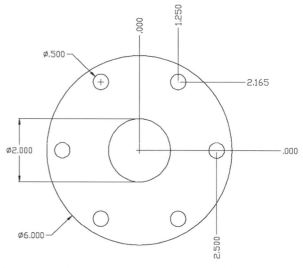

Figure 3–10 Ordinate dimensioning for radial patterns speeds setup.

For turned parts, the best dimensioning method is an ordinate scheme with all the Z axis features dimensioned to the far left side of the feature. Take a look at Figures 3-11 and 3-12 to see the difference.

The image in Figure 3-11 is a jumble of dimensions with the machinist left to sort out by calculation what they need to make the part. The reason for using ordinate dimensions on lathe-turned parts is in how machinists reference their tools. Typically they are touched off on a freshly-faced end which becomes the datum for the rest of the features.

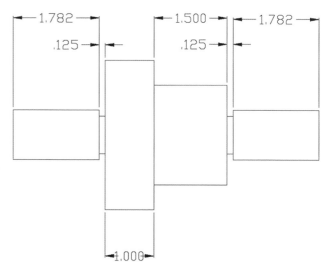

Figure 3–11 Dimensioning not optimized for lathe operator. Math required.

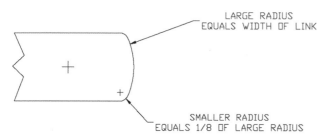

Figure 3–9 A better way to radius links and tabs.

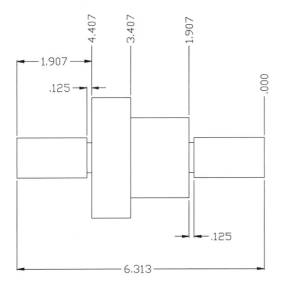

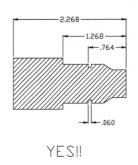

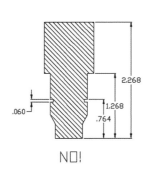

YES!! NO!

Figure 3–12 Dimensioning optimized for the lathe operator. No math required.

Figure 3–13 Drawing of turned part oriented for ease of manufacturing.

The type of dimensioning shown in Figure 3-12 can save time at the machine, with less calculating and more metal removal. For two-sided parts, you can have two datum ends. This style of dimensioning is all quite simple with electronic drafting, so why not give it a try.

3.4 Other Tips

Orient the drawing of turned parts in the same orientation they will be in the machine (Figure 3-13). On the same note, if possible, orient rectangular parts with the long axis running right to left. If you need to pick a datum corner, use the upper left hand corner if it makes no difference to your tolerances and features (Figure 3-14). The second-best choice would be the lower left corner. In general, we like to have the dimensional datum corners butted against our fixed reference surfaces on the machine.

In the modern electronic drawing age, it is a rather simple thing to output a complete drawing for a right-hand, left-hand, or mirror situation (Figures 3-15 and 3-16). Many mistakes are created in the shop because of confusion when working with the "wrong hand view." Every person who handles the drawing after the designer adds the note, "Right hand shown, left hand opposite" is in a position to make an error and waste time.

All must interpret the details of the undrawn configuration. The minor savings in time in the drafting department is eaten up by the first person who has to read the drawing.

Never scale the electronically drawn part to "fit" the title block for printing purposes. Accurate dimensions should always be preserved electronically. If you must scale something, scale the good-for-nothing title block to fit the drawing. Many times these drawings are transmitted electronically and are never printed.

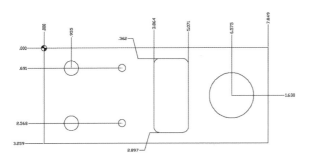

Figure 3–14 Part dimensioned using ordinate dimensions and upper left corner datum.

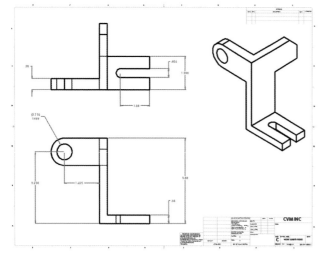

Figure 3–15 Right hand left hand as separate drawings.

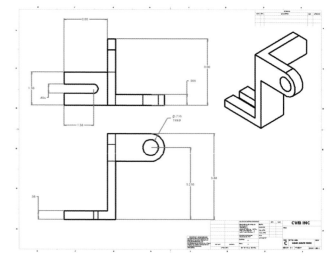

Figure 3–16 Right hand left hand drawings.

Be willing to listen and compromise if it makes sense. On the flip side, explain why you need something a particular way if it's important. A little information never hurt anybody.

Provide thread reliefs on OD and ID threads—typically at least one thread pitch as a bare minimum. You can't use a .015 wide thread relief with an 8TPI thread. The threading tool is 10 times this width just to make the depth correct.

Choose standard fine threads for custom-threaded parts if you have a choice or it doesn't really matter. Fine threads cut quicker than coarse threads. In tough materials, this can make a big difference. Along those same lines for custom threads, pick even numbers if it makes no real difference. On the lathe this makes threading a microscopic bit easier because of the thread chasing dial engagement points.

If you have the mating part for threading, machinists almost always prefer to have it available to double check their work.

I learned the hard way how difficult it is to sit in a drafting room designing parts or entire machines without the benefit of physical things in front of me for assistance.

The next time you get mad at an engineer or designer take a deep breath and have a little empathy. These folks are just as hard working as any shop person.

Office professionals are working with a different collection of information about the project than you are. What may seem like a simple change or solution may have gone through dozens of painstaking iterations in mind-numbing meetings and have much deeper roots than apparent from the shop.

Cost, safety, and yes even political reasons shape the designs we see on the shop floor.

If you can find a good working balance between the shop floor and the office you are on your way to a much more rewarding time spent at your craft.

4.1 Floors
4.2 Light
4.3 Food Areas
4.4 Heating and Cooling
4.5 Workbenches and Tables
4.6 Air Supply
4.7 Raw Material Storage and Handling
4.8 Material Identification and Characteristics
4.9 Safety Equipment
4.10 Tool Crib
4.11 Bench Work
4.12 Filing
4.13 Saws and Sawing
4.14 Rigging and Lifting

Setting Up Your Shop

The condition and maintenance of your metalworking shop have a direct effect on your efficiency and morale. A bright, well lit, clean shop is easier and less mentally demanding than a wet, dripping, bat infested cave. If it looks and smells bad, it probably is.

An organized and well thought out shop space is a pleasure to work in. Having everything at hand and organized is like shopping in your favorite tool store. Retail hardware stores are a good example of what a nifty metalworking shop might look like. Good lighting combined with easy-to-find, well-stocked storage make the work run smoothly and efficiently. Don't underestimate the value of infrastructure improvements on efficiency and profits.

4.1 Floors

Floor colors. Light-colored floors are best for machine shops and areas where fine work is being done. They reflect light and give the shop a cleaner feel. It doesn't hurt that it's easier to find a lost part on a light-colored floor.

Epoxy coating or even special-colored flooring tiles can be used to brighten the shop.

Individual tiles are best because they can be replaced if they get damaged. The tiles also provide a little cushion if you drop a tool or delicate part. Plastic floor tiles in an array of colors are now available—the tiles snap together to make an excellent shop floor.

Welding areas. Welding areas need to have your basic plain concrete floor. Epoxy coatings or paint take a beating when hot metal lands on them. Depending on the class of welding work, paint or tiles may be acceptable.

Steel floors. Figure 4-1 shows a heavy-duty, steel working floor with cast iron platens set at floor level. These platens are used for fixturing and setup of large steel structures.

Smooth finish. A smooth finish on concrete makes sweeping easier. Nor does the finish wear out your boots when you're crawling on hands and knees. Skip the non-slip sand and abrasive material in paint and epoxy coatings. Sample a test section before you commit your floor space into sandpaper. If you need proof, sweep the parking lot for a half hour; see what you think about roughened surfaces as shop floors.

Figure 4–1 Steel working floor.

Wood floors. When it comes to durability and working comfort, wood floors are among the best floors that I have ever seen. The little extra cushion provided by wood makes long hours on the feet and knees easier. However, wood floors are not always practical in the modern, tilt up, concrete shop buildings.

Wood brick floors. In many old time shops, you can still see wood brick floors (Figure 4-2). They are made up of thousands and thousands of 4 × 4 or 4 × 6 blocks cut off three or so inches long planted in the floor end grain up. Some of these floors are 50+ years old and still going.

These floors are made from cheap material and are easy to fix if they get damaged.

Plastic floor tiles. Figure 4-3 shows an interlocking plastic floor tile. These tiles can be snapped together quickly to create a great looking floor in no time.

This type of tile has the advantage of simple installation. Furthermore, you have the ability to take them with you if you move.

Figure 4–2 Wood brick floor.

Figure 4–3 Interlocking plastic floor tile.

Table 4–1 Summary of Floor Features	
Floor colors	Light colors good for fine work.
Floor tiles	Brighten shop; tiles easy to replace
Welding areas	Basic plain concrete
Steel floors	Good for large steel structures
Smooth finish	Easier sweeping; less wear and tear
Wood floors	Extra cushion
Wood brick floors	Inexpensive, durability, easy to replace
Plastic floor tiles	Easy to assemble

4.2 Light

Every shop needs light. Standard warehouse-type lighting just does not cut it. The finer the work, the more light is needed.

If you're planning a shop—regardless of size—and an architect tells you how many lumens per square foot you should have for a machine shop, be sure to upgrade to the next level. The architect may be reading a book that was written shortly after the earth's crust cooled and people worked in caves using mammoth blubber lamps. Things have changed since then.

Adding this extra light will save you in the long run from cobbling in extra fixtures because you don't have enough. Have you ever been in a shop with too much light?

Your model here is the well-lit hardware store or other retail space. The best lighting is equal in intensity from any direction. Not always easy to do, but a good goal.

A bright, well-lit shop is a simple morale building tool.

Modern efficient high intensity fluorescent is best, followed by Mercury vapor lamps; they may need to be mounted high.

Stay away from sodium vapor because is casts a sickly, yellowish light. Mercury vapor lamps take a while to warm up; if you need instant light, fluorescent will be your best bet. In shooting pictures for this chapter, I had trouble with the lighting in one shop that had sodium lights, which required some color correction.

The lights should be mounted fairly high to miss the tops of machines (Figure 4-4). I have seen many bulbs broken in low-hanging fixtures from flying parts and from handling long materials in the work areas.

Task lighting should be easily positioned and, ideally, cool running. Skip the food warming flood lamps. It takes it out of you to have a light that could cook a hamburger beating down on the side of your head all day. Magnetic bases can easily be moved to new positions. Several excellent cool running LED machine task lights are now available.

Figure 4–4 High mount main lighting, task lighting, with added natural lighting.

4.3 Food Areas

Every engine needs fuel and every shop needs a refrigerator and at least one microwave oven. Even if your shop is a one-person operation in your converted garage, you'll want to be able to reheat a cup of coffee or keep cold water handy.

If you have a lot of workers, consider having two or three microwaves. This eliminates long lines and folks popping their stuff in before the lunch bell to beat the logjam for a single microwave.

There is always one person in the shop who has to heat fermented mackerel and sardine casserole to the temperature of the surface of the sun and permanently contaminate the only available oven. If you have two ovens you can designate one for questionable food items and forensic leftovers.

The refrigerator should be cleared out once a month or, better yet, once a week. Stuff tends to accumulate there and be forgotten. Get rid of anything that looks like a failed science project or hides when you open the door.

For larger spaces, you should have a lunch room or break area separate from the actual shop. It's nice to sit down and not hear the CNC machines still running while you're taking a break. A large table promotes camaraderie and boosts morale. Get a couple of marginally work-related magazine subscriptions and leave the magazines in the lunch room.

4.4 Heating and Cooling

Many metalworking shops start out life as warehouse spaces, garages, or even sheds. They are typically cold in the winter and hot in the summer. If you have a choice, insulate and then add dedicated heating and cooling.

Typically these areas need to be kept at different temperatures than office spaces. Be sure they have their own climate controls. I don't think you would get much complaint from anyone if the temp was kept at a year-round 68 degree F. This also keeps your precision measuring equipment closer to their calibration standards.

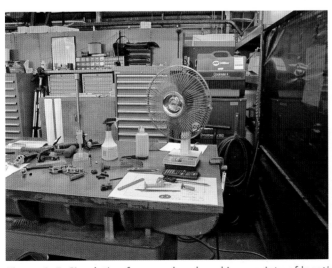

Figure 4–5 Circulating fans can be placed in a variety of locations.

4.5 Workbenches and Tables

Several types of specific workbenches and tables are needed in any shop. All should be of sturdy construction. A workbench that wiggles or is rickety is very annoying. The basic types are basic work and assembly, welding, and mobile. All three have special requirements. Many smaller shops on a budget build their own when they have slack time. (Welding tables are discussed in Chapter 9.)

Basic Work and Assembly Table. This basic table should be rectangular in shape with a maximum width of 36 inches. Much wider than that, and it's hard to reach across without walking all the way around.

Surface and Rim. We surface our general benches with a white board material similar to what is used for dry erase boards (Figure 4-6). It's sometimes called melamine at our local home center. We add a short rim on three sides on some of the benches near the machinery to keep tools and parts from rolling off the back or sides.

Height. The table's height should be 36–40 inches for most folks. Lower gives you back pain after a long day. Generally, higher is for fine work and lower is for heavy work.

Light-colored replaceable tops reflect light and let you see small parts easily. When they get beat up, change them out and they look brand new. Beware of the lower shelf; it's an area that tends to collect junk.

Wheels. I used to believe that the only thing that should be on wheels is the hand truck and the forklift. I have since reversed my thinking completely. For maximum versatility, most workbenches, if not all, should be on wheels—with the exception of heavy welding tables or benches that have machinery attached to them. These tables and benches need to be very stable to do decent work.

Let me clarify this a bit. The wheels need to be swivel with a brake and a swivel lock. If you don't use this type of wheel, skip putting wheels on a workbench. The swivel lock, sometimes called a total lock, is the key ingredient that keeps the workbench from moving around until you want it to.

Having most of your workbenches on wheels allows you to re-configure your workspace quickly for any job that comes in. Such workbenches are easier to move for cleaning purposes and can be moved if more floor space is needed for a large project.

For jobbing shops that never quite know what will crawl through the door, or what the customer trots in, bringing the table to the job is valuable.

Figure 4–6 A rim will keep tools and parts from rolling off the bench.

Figure 4–7 Mobile cart.

Mobile Workbenches and Carts. Smaller than regular workbenches, these are used to move raw material and parts between machines and processes (Figure 4-7). Good wheels are a must for these often overloaded carts. Small is beautiful because of where they are used. They are easy to stow and push around because they hold a couple of hundred pounds at most. We also use furniture dollies for moving heavy plates and boxes around to the different work centers (Figure 4-8). They are cheap and roll well, even with a heavy load. The carpet on the surface keeps your stuff from sliding off and they store relatively flat.

Figure 4–8 Furniture dolly.

4.6 Air Supply

The shop air supply is a critical path in the modern metalworking shop. If your air system goes down, it can affect your entire operation. Many CNC machines require air at a certain pressure and volume to run. A high-quality, properly-maintained system is crucial.

Shop size and machinery dictate the size of the compressor. When choosing your piping system, compressor, and storage tank, be sure to engineer in room and branching possibilities for future expansion. It's easy to install a tee instead of an ell to allow future expansion opportunities.

The sign in Figure 4-9 is not the one anyone wants to see on the first day of a new welding job! I hope they drained the water from the air system this week….

Consider multiple smaller, strategically located air tanks instead of one single large storage tank. Often it's easier to find precious floor space for smaller tanks.

The distribution network should be steel or copper pipe. PVC works and is safe under most settings, but has the cheapskate rookie look to it. OSHA says anything below eight feet off the floor needs to be metal, so you might as well make the whole thing out of metal. Copper gets my nod of approval for corrosion resistance and low leak potential.

If you are unsure whether you need a valve in an air system, my motto is, when in doubt, add a valve. Each air line drop and machine air supply line should have its own isolation valve.

Depending on your shop layout, you should be able to isolate areas for repairs, maintenance, or expansion without having to shut down the entire air distribution system.

Figure 4–9 Air supply is essential to the shop.

Simple drain cocks should be in the bottom of every drop in the shop to bleed accumulated water (Figure 4-10).

Air nozzles dedicated to machines should be plumbed directly to the hoses without quick disconnects (Figure 4-11). This prevents not only the unwanted removal and inevitable wandering of the air nozzle, but also the time wasted hunting one down.

For whatever reason, some drops seem to collect more water than others. It's great if you have a chiller dryer on the compressor. But very few air systems I have used were completely free of water. Therefore, provide a way to get the water out and use them diligently.

Assign the duty of draining the compressor to one of the apprentices on a regular basis, typically at least three times a month.

If the compressor is used heavily, invest in an automatic drain.

The air supply to the air nozzle should be regulated to a lower pressure than the main shop air. It cuts down on the raw noise and is safer against

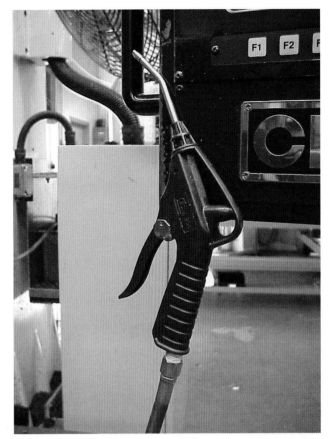

Figure 4–11 Dedicated air nozzle.

bare skin. Over time, this noise on unprotected ears will cause hearing damage.

The hose for a dedicated nozzle should be optimized for length. A long coil of unused hose lying on the floor collects chips, coolant residue, and junk. It's hard to clean around, let alone the leg-snaring abilities of simple rubber hose. For air blower nozzles, a small inside hose diameter (1/4) makes for a lighter, more flexible hose. I have never had a problem with flow for an air nozzle using this size.

Figure 4–10 Simple drain cock.

Air Tool Supply Lines. These lines are sized for each tool. Use your largest air hog to determine inside hose diameter (Figure 4-12). If you're doing delicate work with small tools, you can make up a short whip of hose with a smaller inside diameter that connects to the main hose. Making these six feet long at maximum keeps the large main supply hose from dragging your arm down all day long. On a couple of my tools, I have plumbed the whip right into the tool body to eliminate the bulky connector or swivel joint.

Hose reels are great for neat clean storage of air lines (Figure 4-13). Put two at opposite ends of the shop and have them slightly overlap in length with the hose fully extended. Don't forget to put a universal female coupler on the end. Make sure that it sits higher than the tallest guy in the shop's forehead, unless you like the knurled forehead look.

Don't let the hose slide through your bare fingers when retracting the hose reel. Razor sharp chips get stuck in the surface of the hose and will slice your hand.

Instead, keep your hand on the quick disconnect at the end. Pain awaits the daydreamer with a bad bashing. Be sure to include a valve right before the hose reel so you can service the reel.

Don't drive forklifts and other heavy-wheeled machinery over light-duty air lines. Otherwise, you will shorten their life considerably by driving chips and scrap metal into the surface. I got yelled at for this many times as a teenager in the shop, so now I get to pass it on.

Air tool oil should be readily available near air supply points. This accessible location encourages the lubrication of expensive pneumatic tools. Actually it encourages a convenient storage place to keep the **empty** bottle of air tool oil, but that's another story.

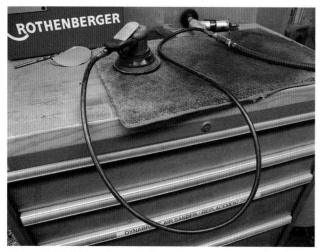

Figure 4-12 Air tool supply line whip.

Figure 4–13 Hose reels help store air lines.

Provide simple professional looking hooks for storing air hose extensions (Figure 4-14). A similar hook works well for getting extension cords off the floor. Nails and spikes are for carpenters and other woodchucks, your hoses and cords will thank you with an extended life.

Universal female quick disconnects (Figure 4-15) fit several of the most common types of male plugs encountered in air systems. They save time wasted hunting down the correct fittings or adapters. It seems like every shop independently decided to use a fitting diametrically opposed to every other shop's fitting. I have a little box in my toolbox with every adapter under the sun to prove this.

Figure 4–14 Hooks help store air hose extensions.

Figure 4–15 Universal female quick disconnects.

4.7 Raw Material Storage and Handling

Every shop must store raw materials. The trick is to store them so you can get to the piece you want with the minimum of human effort. Labor is typically the most expensive component of any job. Anything that shortens the labor path during a job will have a short payback period.

Cutting bars. Cut 20-foot bars in half for easier handling. It depends on the type of work a shop does, but rarely are full 20-foot bars utilized without cutting. In small shops, 20-foot lengths of heavy materials can be extremely difficult to handle; they can tie up expensive labor to cut off a short length.

This challenge goes on and on until the material gets short enough to be handled efficiently.

It holds true especially for machine shops using short fat chubs.

The time to cut the bar is when it comes in the door from the supplier. I don't know about you, but I'm too old to buck 12-foot lengths of 3-inch round out of the bottom of the rack.

Using carts. Figure 4-16 shows a cart we made to move bar stock around the shop. It is the correct height for the horizontal band saw so material can be pushed right into the saw as soon as it's unloaded.

The support tubes are set up so you can load and unload it with a forklift right off the delivery truck.

Organizing materials. Store materials by length, for example, full bars, partial bars, and stubs.

Most people are lazy and will try to use the material with the easiest access before pulling out the full length materials.

Figure 4–16 A cart helps when moving bar stock.

Systems that find the optimum stock in the speediest manner shorten the job. This is proven when we filter the racks and bins (Figure 4-17). The size of the keepers get smaller and smaller until they just disappear.

Weeding materials. Weed the material storage once in a while. This job is a good one for apprentices and helpers. They will get acquainted with the shop's raw material inventory. This task provides a great way to give them direct contact with many different material types, sizes, and forms they will need to know about later as their careers progress.

Beware of packrats. Machinists and welders are natural scrounges and packrats.

Want proof? Look under the workbenches and in all the nooks and crannies in their work areas. Some materials and leftovers really need to be scrapped for running an efficient shop.

Get rid of materials that cannot be easily identified. If you don't know what it is, how can you use it?

If you must be a packrat and save it, mark it unknown. "When in doubt, toss it out."

Certain form factors of scrap and material drops are virtually useless. Long thin strips of sheet metal and triangles that are cut off from almost anything are two examples that can hit the scrap bin right out of the gate.

Almost all circles and discs should be saved. Unless your processes generate a large quantity of disc-shaped parts, these should be saved.

Storing materials. Store flats on edge in the vertical orientation (Figure 4-18). Graduate them by width to make inventory and removal easier. Otherwise, the one you want is always on the bottom.

Store your most accessed materials at waist height. I always hate it when I get my eye core sampled by a piece of tubing when I'm reaching over my head for something that should be lower.

Metal supplier color codes are unreliable. There is no real color standardization in the metal supply industry. These should only be used as a

Figure 4–17 Racks and bins provide additional storage.

Figure 4–18 A good method for storing flats.

generalization or identification when combined with another method.

Small rounds should be stored in tubes so they don't slip between the rack dividers (Figure 4-19). You can also form simple sheet metal trays for the slots that contain small materials (Figure 4-20). This keeps more flexible materials from drooping and missing the rack supports.

Figure 4–19 Use tubes to store small rounds.

Figure 4–20 Storing other small materials.

Identifying. Material identification is a challenge to maintain in a job shop. Each time a piece of material is cut, there is a chance to lose the identity trail (Figure 4-21). It's every person's responsibility to make sure the identity chain does not get broken.

Keep marking tools readily available in areas where materials are stored and rough cut. Put the identifying marks on the end surfaces of bars. Be sure to mark both ends. Engrave or mechanically stamp the material type directly on the bar end (Figure 4-22).

Tags fall off; ink smears. Tape or labels are next to useless. Other material sliding in the rack can obliterate the identification. The sticky labeling goo also has to be removed by somebody on the payroll. I hate label goo on my consumer packaging, so why would I want it on my raw materials? When you mark the end surfaces of material, it fits in collets and vises; the first cut in the machine takes care of removing the engraving.

Don't use obscure internal company terminology or secret codes for common materials. Use the same terms suppliers use. For certified materials, include the purchase order number generated when the material was bought; this helps you track copies of test reports and conformance information.

Figure 4–21 Identify your materials.

Figure 4–22 Indicate the material type.

4.8 Material Identification and Characteristics

Machinists and metalworkers need materials from which to make things. There are literally hundreds of different materials that are considered common nowadays.

When you think about all the different kinds of operations that a piece of material might be subjected to during its fabrication, the knowledge required to keep track of the different characteristics is daunting. There is not one go-to place to learn these things—unfortunately, most are learned the hard way.

At one place I worked, I was subjected to six distinctly different materials during my first two days. Each had a specific set of do's and don'ts.

For the modern metalworker, a solid knowledge of materials and their qualities goes with the trade. When you are given an unfamiliar material to work with, take a quick minute to look up or at least ask about the common characteristics and problems. Some materials are so sensitive to common shop chemicals and substances that they can be permanently damaged without you doing anything intentionally. "When in doubt, check it out."

The ability to identify different materials at a glance by their look, feel, and mechanical qualities takes time to develop, but the practice necessary to learn is well worth the effort.

Any given shop will have a cross-section of different materials—several dozen at least—that the metalworker meets. How often have you picked up some unmarked material and wondered exactly what it was? These quick, shop expedient methods can narrow the field of doubt.

Table 4-2 Identifying Materials
Basic Characteristics
Color
Magnetism
Density
Hardness
Chip forming
Thermal conductivity
Mechanical Condition
Mechanical
Modulus of elasticity
Yield strength
Toughness

Our hands and eyes are very sensitive comparative instruments. When trained, they can discern minute differences in color, thermal conductivity, ductility, density, and modulus. These are not absolute methods, but will help narrow the field and tip the scales in your favor when examining and choosing different materials. The more you know about the different materials you work with, the more you will be attuned to each of their special qualities.

This intimate knowledge of materials is especially useful when building or designing parts with different functional requirements.

Making comparisons with a piece of known material in hand will improve the accuracy of your observations. If you think something is 7075 aluminum, have a piece of known 7075 while checking the unknown material.

Don't rely on any one characteristic. Use as many as you can compare to remove doubt.

Color. Aluminum and steel have distinctly different colors. Both are silver, but aluminum has a slight bluish tinge that steel lacks.

The color difference between steel and stainless is less obvious, with stainless having a slightly more silver color and steel more toward grey. Copper is redder than brasses or bronzes and brass is more yellow than most bronzes.

Magnetism. Small magnets on pocket screwdrivers (Figures 4-23 and 4-24) can tell you a lot when you're hip deep in the metal rack or you see a funny looking dowel pin in the box of stainless pins.

I always have a cheap magnetic screwdriver in my apron pocket. It's my combination pry bar and metal identifier.

Density. Many materials can be identified by their density or lack of it. 7075 aluminum is noticeably heavier than 6061 for an equal-sized part, even at only 3% heavier. This density, combined with higher hardness, makes 7075 stand out from the more common 6061.

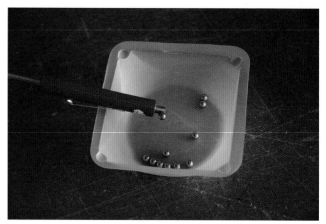

Figure 4–23 Using a small magnet.

Figure 4–24 Pocket screwdrivers have many uses.

Figure 4–25 Distinguishing between materials of different hardness.

Chip forming. Each material has a characteristic chip formation. For example, when it is compared to either 304 or 316, 303 stainless tends to short chip.

Delrin makes short chips that smell different than ABS or polyethylene during cutting.

Most plastics have distinctly different smells when a small sliver of the material is heated. Caution: some fumes from heated plastics are detrimental to your health.

Thermal conductivity. Aluminum feels warmer to the touch than stainless of similar finish.

Mechanical condition. Harder material with higher tensile strength will ring at a higher pitch when dropped or tapped with a tool.

Hardness. Different aluminum types can be detected by the differences in hardness. This test also works well for many plastics. For example, Delrin is harder than ABS or polyethylene. Acrylic is harder than polycarbonate. A simple test with an automatic center punch (Figure 4-25) can distinguish between two materials of different hardness.

In the example shown in Figure 4-26, the sample on the left is 7075 Aluminum and the sample on the right is 6061.

The center punch mark is slightly larger in the 6061 sample than it is in the 7075 sample. This comparison is what we would expect to see in this case, knowing what we do about the characteristics of the two materials.

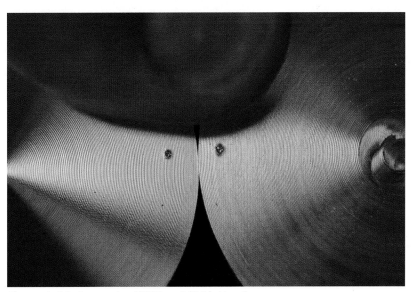

Figure 4–26 Comparing 7075 and 6061 Aluminum.

Get your hands on an armload of metal and plastic supplier catalogs (Figures 4-27 and 4-28). These provide a wealth of useful information on material properties and processing. It's easy to flip back and forth in these great little books comparing the different qualities. Once you have some experience in the shop with some of the materials, you can easily correlate the observed differences with the book data.

I keep a big binder I call my "Big Book of Knowledge." It has all the

Figure 4–27 Metal and plastic supplier catalogs.

data sheets and information I have had to look up in the course of doing my job. When I get nice information on, say, the factory flatness standards of stainless sheet metal or the corrosion qualities of Titanium in nacho cheese sauce, I put it in the book. It has been extremely handy over the years because it relates directly to personal experiences.

Figure 4–28 Supplier catalogs provide valuable information.

Elongation. Elongation numbers gives us a clue about the formability and ductility of a material. The higher the percentage, the farther it can be worked.

Modulus of elasticity. This measure tells us about a material's stiffness. The higher the number, the stiffer the material.

Stiffness and strength are two very different qualities. Don't get the two mixed up.

Yield strength. This measure shows the value we use to determine the upper end of basic strength differences between materials.

Ultimate strength is not as useful as a design parameter because most materials have already permanently deformed or yielded long before they see the ultimate strength.

These numbers just give us a relative comparison of strengths. Yield strength shows us at what point the material will be dimensionally altered.

Toughness. This indicator measures a material's ability to absorb energy. Impact and notch test numbers give us a way to compare different materials. Higher numbers mean tougher material. They also mean more sweat if you have to apply human energy.

Sheets and Plates. Sheet is anything under .188 thick. Anything thicker is called plate.

Sheets and plates have bow (parallel to the rolling) and camber (perpendicular to rolling) tolerances that would surprise you.

Most of the time, these materials come in tolerances better than what is allowed, but sometime they don't. These specifications are available from your suppliers.

If they don't want to give you the information, take your business elsewhere. Use this valuable information for your real world design efforts and tolerance analysis.

Cast aluminum tooling plate sold under various brand names is not ground flat. It is cast against a ground surface, which gives it that ground look that sells more plate (Figure 4-29).

This material is not particularly flat by machinist flatness standards. First get the specification sheet from your aluminum suppliers.

Then get ready for a surprise when you see how much flatness deviation they graciously allow themselves.

Keep in mind that many of the cast varieties of tooling plate offered are not repairable with welding.

Figure 4–29 Cast aluminum tooling plate.

Figure 4–30 PVC covering helps to protect finishes.

There are only one or two types of tooling plate that I know where the welding is even recommended. Screw ups can be hard or impossible to repair. The major redeeming factor is that large amounts of metal removal have little effect on geometry.

Aluminum sheet is specified by decimal thickness, not gauge number. The salespeople generally overlook this common oversight—they want to sell you material and not make you look ignorant. Order your sheet materials with PVC covering to protect the finishes (Figure 4-30). The small incremental cost is repaid in less finishing work.

For finishes produced in the shop, we sometimes protect the surface with clear shelf paper (Figure 4-31). It is readily available at any hardware store and will protect the finishes you create as they move through the shop. We recently started using a similar protective sheet material called Surface Armour. It is available in many different levels of tack, depending on the material you want to protect and how long you need to protect it. It is available in rolls in any width you need. See the suppliers' index for contact information.

Figure 4–31 Clear shelf paper also protects finishes.

Carbon steel and stainless steel sheet of the same gauge number have different thicknesses. To avoid confusion, call out the sheet metal thickness as a decimal dimension with a tolerance on a drawing.

Include an appropriate tolerance to the thickness to encompass the manufacturer's gauge tolerance. Beware: plastics typically have thickness and diameter tolerances of +/– 10%. These variances catch many people by surprise because of their large size.

All sheet materials have a thickness tolerance from the manufacturer. Be aware of this tolerance to avoid mistakes and to use to your advantage. This tolerance shows itself when doing precision forming where the thickness effects the forming operation.

Sheet materials are usually thicker in the center of the sheet parallel to the rolling direction than near the edges.

Sheets are always slightly oversize in length and width. You can never count on the accuracy of the sheet width or length. There is a reason they sell materials by the pound—all the extra weight goes on your bill. If you really need an exact width, plan on re-squaring the material.

Galvanized sheet is hot rolled sheet that has been pickled in acid to remove the mill scale before plating. It is softer than cold rolled sheet of equivalent thickness. It's handy to keep around for durable templates and things you don't want to paint.

Table 4.3 Sheetmetal Gauge vs. Actual Thickness

Material	Sheet Gauge Number	Actual Decimal Thickness
Steel	11ga	.1136″ min to .1256″ max
Stainless Steel	11ga	.125 +/– .007″
Aluminum	11ga	.090 +/– .005″

Material Lengths. Depending on material type and form, bar materials come in many lengths. When calculating material needs, understand how the material will arrive so proper allowances for yields can be figured. I do not understand why suppliers hold these odd differences. The industry should standardize this.

Steel pipe and tubing come in 21-foot random lengths. Okay, I get this—twenty feet plus a little. Extruded aluminum bar shapes come in 12-foot lengths. Not sure about this one. Maybe is has to do with droop when it exits the extrusion die.

Extruded aluminum tubing comes in 20-foot lengths. There goes my theory on droop.

Aluminum angle comes in 24-foot lengths. ***This*** is the pipe to beat!

Cold rolled steel bars come in 12-foot lengths. Precision cold rolling might explain the need for shorter bars. But many have sheared ends so they would appear to be rolled as longer lengths.

Hot rolled steel bars come in 20-foot lengths. Why not 21 feet? Stainless steel bars come in 12-foot lengths, depending on whether the bars are true flat bars or Gauer bars, which are slit from coil.

If you are calculating yields for single part lengths, this is all straight forward. Part length plus a little for cutting; divide this into the bar length. An assembly like a large machine frame with many different cut lengths can get a little tricky. With expensive or hard-to-procure materials, a few minutes spent looking at a decent cut list can save thousands of dollars. This is best handled by the person who normally does the cutting.

Pipe Sizes. All true pipe of a given size will have the same outside dimension. The inside diameter changes with the pipe thickness schedule. This is a carryover from the old days when pipe connections were threaded. You only need one size threading die for each pipe size. Now, all it does is confuse purchasing agents who think everything round with a hole in it is pipe.

A couple of pipe sizes that make great telescoping assemblies are 1-inch schedule 40 and 1-1/4 schedule 40. These make great adjustable stands (Figure 4-32). For heavier duty stands, 2-inch schedule 80 and 1-1/2 schedule 80 make good telescopes also.

Square tube telescopes are more difficult to get right. Internal weld seams and lack of size choices limit this. An alternative is to add material to the outside of the inner tube to make a telescope. Plastic strips can be attached to provide a bearing surface for the telescope.

Special Materials. Several special materials are worth mentioning here; you should become acquainted with them if you don't already know them.

Aluminum bronzes. These super-strong bronze alloys are a near perfect match for sliding contact with most stainless steels. Drilling, tapping, and reaming can sometimes be a minor pain. Tensile strengths can go higher than steels with some alloys.

17-4 Ph Stainless steel. This versatile precipitation hardening stainless has similar corrosion resistance to type 304, but can be hardened with a simple low temperature heat treat to RC44. It welds with normal processes without fuss. It is very stable in heat treatment and can be machined easily after heat treatment. Unlike some of the other precipitation hardening stainless steels, this alloy is readily available in many forms.

Figure 4–32 Pipes can be assembled to make an adjustable stand.

Steel by the trade name Stress-proof, otherwise known as 1144. This steel is an easy-to-machine, tough, strong steel. Over 100KSI yield strength as delivered, it can be further heat treated. It is a favorite of the screw machine industry. Its only real problem is that it not recommended for welding.

8620. This alloy is another strong easy-to-machine steel. It can be purchased most commonly in rounds for turning work. It is easily welded by all common welding processes. The carburizing heat treat can

Figure 4–33 8620 is an easy-to-machine steel.

be easily controlled to produce a deep hard case with a tough core, as seen in Figures 4-33 and 4-34.

Figure 4–34 Producing a hard case with a tough core.

Plastic Tips. When working with plastics, beware of those characteristics that make some operations difficult. Most plastics expand much more per unit length per degree than metals—up to 10x greater.

Keep heat out of plastic parts. Plastics soften and melt easily under the friction of cutting tools. Try to use sharp tools and coolant when possible to reduce melting; remove the heat generated by cutting.

Plastics conduct and shed heat much more slowly than metals and can be damaged by localized heating. Slower spindle speeds and higher chip loads can help if you are having a problem.

The difference between the two demonstration drilled holes in Figure 4-35 is that the better looking one was peck drilled at a much slower spindle speed.

Both holes used the same feed rate. The cycle time was marginally shorter for the bad hole, but who cares? It's a lousy hole.

Table 4.4 shows cutting speed and feed rate start points for drilling some common engineering plastics with high speed drills. Increase feed rate per revolution as drill size increases. Use coolant if possible.

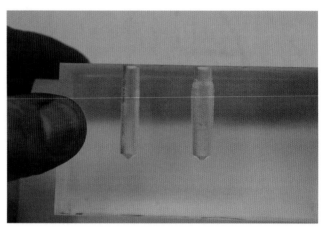

Figure 4–35 Spindle speeds can affect the quality of drilled holes.

Table 4.4 Cutting Speed and Feed Rate Start Points

Material	Cutting Speed (Drilling)	Feed Rate
Acetal (Delrin)	250–500 FPM	.007–.015″/Rev
Polycarbonate (Lexan)	150–250 FPM	.007–.015″/Rev
Acrylic (Lucite) (Plexiglas)	150–200 FPM	.007–.015″/Rev

Controlling Chip Loads. Chip load = Feed rate (inches per min)/(RPM × Number of flutes)

Example:

5 inches per minute / (1000RPM × 2 Flutes) =.0025 per flute chip load

To increase your chip load you can decrease the RPM, increase the feedrate, or use fewer cutting flutes.

To decrease your chip load you can increase the RPM, decrease the feed rate, or increase the number of cutting flutes.

For intricate parts with tight dimensional requirements, you may want to try stress relieving your plastics. Most heat treat shops can handle these requests.

Be sure to ask your plastic supplier for the proper temperature and time profiles for your specific plastic.

This task can be done before machining and fabricating, or as an intermediate step prior to final machining. You can do this work yourself in an oven with a good temperature controller and timer (Figures 4-36 and 4-37).

Typically, you heat to just below the heat softening temperature and hold for a time related to the part volume. Then reduce in controlled steps until reaching ambient.

Here is a typical stress relief cycle for acrylic. Heat the part to 180°F over 2 hours. Hold at 180°F for 30 minutes per each 1/4 inch of part thickness. Cool down at 50°F per hour until you reach room temperature. This will help eliminate any residual machined in stresses.

Figure 4–36 You can stress relieve plastics in an oven.

Figure 4–37 The over should have a good temperature controller and timer.

4.9 Safety Equipment

Living is a dangerous occupation—just look at how many dead people there are. Metalworking is no more dangerous than any other human occupation; its hazards are just different.

I believe that safety is a situational awareness issue as opposed to an equipment one. Just because you have protective equipment and guards in place does not assure any particular level of safety. On the same note, experience and training can help, but do not guarantee safety either.

If you never get in a car, it's less likely you will be in a car accident. If you never stick your arm in a log chipper, you will most likely never have it removed by one.

Two kinds of hazards. There are two kinds of hazards. The first are the ones you can just plain avoid. And the others are hazards that can seek you out.

The story on the next page is an example of a hazard that seeks you out deliberately. So what can you do about these two types of hazards?

Avoidance vs. awareness. Avoidance is pretty effective, but doesn't get much work done. We cannot avoid all the hazards we face in our trade. Therefore, the best form of protection we can use is awareness.

It's hard to sneak up behind a karate grand master. We must have this same situational awareness to the hazards of our trade. We are the prey; the hazards are the predators. If something wants to eat you for lunch, you should pay close attention to what that thing is doing at all times.

I have taken quite a few folks to the clinic to get stitched up because they let the predator have a little bite. The hazards that seek you out are tougher to guard against.

Hardhats, glasses, earplugs, and seatbelts are examples of things we can use to mitigate the seeking variety of hazards in our chosen trade. These and others will save you a lot of grief later in life if you use them diligently. What are the most common traits you have seen with career metalworkers? Two stand out to me. The majority of the old timers I have met are half deaf and three-quarters blind, and all have bad knees. I can't help you with the knee problem, but the ears and eyes can be saved.

Ear plugs. The constant background noise in a metal shop destroys the hearing of most people after years of exposure. Without a doubt, unregulated blow guns and the whine of a geared engine lathe will detrimentally affect your hearing.

Earplugs are a hassle to get used to, but once you get past that break in period, you won't be able to work without them (Figure 4-38).

The "roll up, throw away" ear plugs seem to be the easiest to wear and have a decent noise reduction rating. Skip the earplug leash. If you take the plugs out of your ears and put them back in more than once, they are grubby from your hands or ear wax. The string gets snagged on every protrusion in the shop, yanking the plugs right out when you least appreciate it. If I have to remove earplugs frequently, I switch to ear muffs.

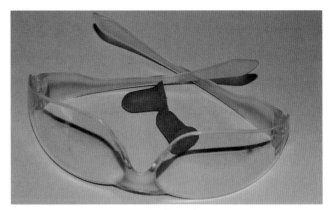

Figure 4–38 Ear plugs and safety glasses.

Hazards at the Garage Sale!!!

I was at a garage sale many years ago—one of those once-in-a-lifetime sales where a retired metalworker was selling off his junk pile. A friend and I spent the better part of the day rummaging around in the various rooms and sheds in a machinery-drunken stupor. We each had a pile of pirate treasure that grew steadily. We were thinking about setting up our tent…and then we struck gold!

Late afternoon, a neighbor asked the owner to help him with some metal-related project. (Later I saw it was a mower blade that needed repair.) I suppose the owner was the Mr. Fixit of the area. This neighbor read the writing on the wall, and figured getting help was now of never!

My friend and I were rummaging about fifty feet away. The owner set up the mower blade in a vise attached to a bench that he dragged out of the basement. Out of the corner of my awareness, I heard him hammering on the errant blade clamped in the vise, but I paid it no real heed.

After a vicious round of hammering, I found a real gem in the pile. At that moment, I stood up and raised the bauble so my buddy could see my find and the gloat on my face. Bing! One more hit with the hammer—and a sudden pain at the corner of my unshielded eye. Apparently Mr. Fixit was using a cold chisel to chip the heads off several steel rivets holding the mower blade to the hub. A chunk of rivet clipped me in the corner of my right eye. I was extremely lucky; the damage was only a small nick and a watering eye for a half hour.

Safety glasses. In any shop that does a significant amount of welding, glasses of some sort are a must (Figure 4-38). Stray UV light eats away at our peripheral vision of an unprotected set of eyes until your vision is ruined. Clear plastic safety glasses are effective against UV light. 99% of the harmful rays are reflected off clear plastic. **Be sure your safety glasses have side shields.** This is the area where our peripheral vision is attacked by the harmful UV light from arc welding.

Scotch Super 33 Electrical Tape. This is the best industrial field dressing for cuts and scrapes. This nice snug wrap (Figure 4-39) looks more painted on than wrapped.

Under normal combat conditions the prepared shop veteran expects a certain number of cuts and bleeds.

A quick wrap of electrical tape keeps blood off the work until you can clean and dress the boo boo. Tape conforms to your fingers' curves and stretches as you move. Unlike bandages, it doesn't leave sticky residue which you have to remove with acetone or some other nasty solvent. Try keeping a normal bandage on with your

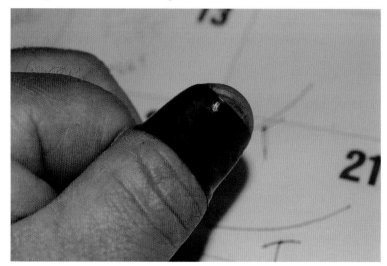

Figure 4–39 Scotch Super 33 electrical tape provides a snug wrap.

hands in coolant all day. I always keep a roll of electrical tape in my toolbox.

Uncle Bill's Sliver Tweezers. Whoever designed these tweezers got it right (Figure 4-40). They taper not to a deadly sharp point, but a small flattened point that meets with micrometer-like precision. Micro-fine metal splinters don't have a chance with this tool. And yes, I really did stick a metal splinter in my finger for your enjoyment (Figure 4-41)!

Figure 4–40 Uncle Bill's sliver tweezers.

Figure 4–41 What I did for my readers!

4.10 Tool Crib

My old teacher Doug, when asked him what tools I should bring into work, told me, "Bring everything. Then as you see how the work goes, you can take the stuff you're not using home" (Figure 4-42).

He worked out of a small sheet metal toolbox that he had bent up at some point. The tools he had in the box were a collection of new and old. Several years before somebody had broken into the shop where we worked and stolen some tools. One of the boxes they took was Doug's.

The part that was slightly funny was that these guys came in through the roof and then broke a window to get out. When the cops were there looking around and making their reports, Doug got one of his tools back. Apparently one of the tool thieves had used his 12-inch crescent wrench to knock the windows out. The cop handed it to Doug who then asked, "Did you check this for fingerprints?" to which the cop replied, "Oh. I guess not."

Doug extended the generous offer that I could borrow any of his tools so long as I was still buying tools myself.

This is a basic ground rule and a great way to foster tool appreciation. If I stopped buying tools, there would no need for me to borrow any of his. You can judge a person's commitment to their trade by the investment they make in their tools.

If you borrow a tool, be sure to return it promptly. It is a major infraction in the metalworking shop to return borrowed tools lazily. It's very disrespectful. The white collar equivalent would be borrowing an engineer's toothbrush or pocket protector and failing to return it.

I have found that engineers and scientists tend to view hand tools as objects instead of treasured personal possessions. Don't make the mistake and create an enemy by not returning a borrowed tool enthusiastically. I've seen some people pretty angry over a 1/8 allen wrench.

Buy tools of the highest quality you can afford. If you're planning to make your living with them, the initial painful cost is amortized over your lifetime. Besides you are directly supporting somebody else who thinks enough of fine tools to actually make them. Any dedicated craftsman appreciates and uses fine tools as part of his or her trade.

Figure 4–42 Doug said, "Bring everything!" So I did.

Part of the fun of doing the work is using the tools. It's not much fun if you use a tool everyday, but regret buying that model or make. There is a deep satisfaction attached to working with your hands. Fine tools go right along with that feeling.

Some people argue that less expensive tools give identical performance. If that were true, expensive toolmakers would be out of business. When skill levels rise, differences become more evident. Some differences are admittedly small, but I think the additional cost is well justified in most cases.

I admit the reverse is also true. I bought a pair of the most expensive European sheet metal snips with carbide cutting edges and high expectations on their performance. For the $150 I spent for a single pair of snips, I expected them to make lunch for me after they were done cutting. On a thrifty · friend's recommendation, I also bought a pair of basic $20 snips. They ran circles around the high dollar ones!

Like carpenters and their tool belts, all efficient metalworkers need a few crucial tools with them as they're moving about the shop. The class of work the shop determines which tools are necessary.

Police officers could not do their jobs very well if they had to make a trip back to their cars every time they needed something.

The same is true for metalworkers. It would be great if you were never more than a step from your toolbox anywhere in the shop. Everything you need would be a short reach away. Countless hours of hunting and using the wrong tool for the job would be saved.

Over the years I have tried different combinations and finally condensed it down to the bare minimum of tools I keep on my person (Figure 4-43). Having these tools at your fingertips at all times saves an incredible amount of time in a given day.

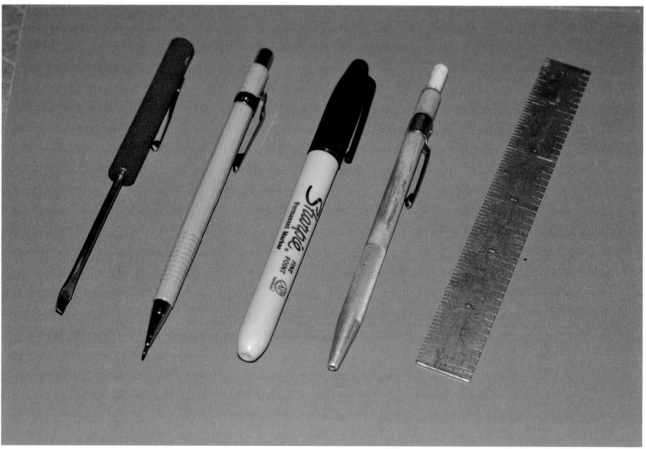

Figure 4–43 Essential tools of the trade.

Tape measure. Work with a 10-foot minimum length. You can remove the belt clip if you prefer to carry it in your pocket. If you can't measure at a moment's notice, you will not measure enough and more mistakes will be made as a result. A very simple way to avoid mistakes and increase efficiency is to carry your tape measure and use it often.

Calibrate your tape measure frequently. Check it against high quality rulers or combination square blades (Figure 4-44). Bend the hook to adjust the zero.

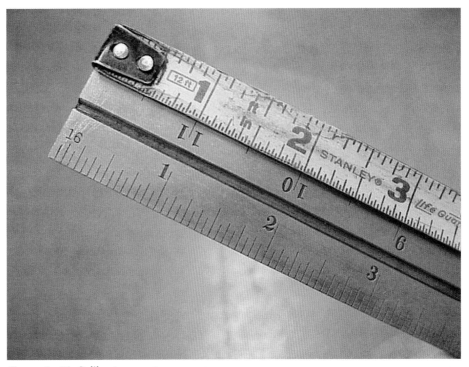

Figure 4–44 Calibrate your tape measure.

Six inch ruler. This ruler is your best friend—well maybe second best. It stirs coffee and double checks a zillion things every day. Skip really fine division rulers. For measuring closer, there are better tools. Know the thickness of your favorite rule. It's as a field expedient feeler gage. The one in Figure 4-45 is .015.

Figure 4–45 A ruler can serve as a feeler gage.

Retractable point scriber. This scriber (Figure 4-46) won't poke a hole in your shirt or apron pocket. A carbide tip is a must if you do any work on stainless steel.

Sharpie marker. These markers can be black or blue for light-colored materials; silver metallic for dark materials; and red for junk parts, screw-ups, corrections.

I really wish I had invented this product. I have these all over the shop—I don't have to walk more than three paces to lay my hand on one in the building.

Small flat blade screwdriver with magnet. This tool is your pry bar, pick,

Figure 4–46 Retractable point scriber.

and a handy screwdriver (Figure 4-47). I like ones that salesmen hand out for gifts. This little screwdriver is handy all around the shop: aligning sheet metal edges prior to welding and loosening the lock screw that holds the punch and die in my Whitney hand punch. The small magnet in the end is important also. You can use it to quickly differentiate between several types of common materials. Don't be afraid to abuse this tool. Think of it as consumable. It will keep you from using your good tools for bad jobs.

An instrument for writing on paper. I like mechanical pencils because they never need to be sharpened. Use thicker .9 mm lead for heavy handed Cro-Magnons. Good written communication is important to success. Unlike ballpoint pens, pencils write on vertical and overhead surfaces. Pencils are king out in the shop. The thin extendable tip of a mechanical pencil will reach into small template holes and otherwise go where a wood pencil can't.

Figure 4–47 A screwdriver, magnet, pry bar, and pick—all in one!

4.11 Bench Work

I love working at the bench. Having all your parts and tools at your fingertips along with a steaming cup of coffee is a truly pleasurable thing. One of my most favorite jobs at the bench is to design and build a tool or small machine on the fly. I use the welding table as my chalkboard to develop ideas. Hand fitting and assembling one of your own designs and seeing it come to life on the bench is extremely satisfying.

A large portion of the work a machinist or welder does centers around a workbench of some sort. There is always something to take apart test fit or clamp down. When you're working at the bench, having a few tools at your fingertips makes for efficiency.

Bench Vise. Without a decent vise, a shop is not a shop. If there's anything that needs to be high quality, this is it.

Once we were building a feed auger for a briquette press. The shop had one of those inexpensive offshore vises that I call ten footers. They look great from ten feet away, but when you get up close you can see all the Bondo used to fill the casting "Irregularities."

We were using this vise to squeeze the auger flighting down onto its center shaft to get some tack welds in place. The vise failed as we worked on this screw, and made the job that more difficult. The boss that the handle slides through buried itself in the Bondo-reinforced front surface, knocking the clamping force in half.

We responded by cranking it even tighter. Something had to give; it was the cheap vise. I was so frustrated I made a call and ordered a new Wilton vise for $350 on the spot. I ended up in a little hot water with the boss, but we still have that vise in the shop and the cheap one has probably been recycled into garden gnomes by now.

Skip the swivel base in the fabrication shop (Figure 4-48). You want the vises used in these areas to be rock solid and not move at the wrong time. It's tempting to get the swivel base, but I guarantee it will let you down. You can't lock it down tight enough with the little handles they put on the clamps.

For most classes of work, use a maximum of four-inch jaw width. Larger than this and the vise becomes cumbersome to use. Try spinning the handle on a six- or eight-in-vise rapidly to open it while you're trying to position a red hot part. If the handle doesn't knock your teeth out, count yourself lucky.

Figure 4–48 Avoid swivel bases whenever possible.

Figure 4–49 Three-point contact adds stability.

Figure 4–50 Using copper jaws as a heat sink.

Clamp large diameter rounds below the vise jaws. Three-point contact makes it more stable (Figure 4-49).

Install copper jaws in all the bench vises in the shop. Everybody puts protective covers over the serrated jaws anyway, so why not eliminate the marking problem? The lame flip-floppy aftermarket covers always fit poorly, which makes it a pain to grip a small delicate part. Name one thing that it would be acceptable if it has big teeth marks in it.

Figure 4–51 Mounting a vise for use with a long part.

It's a lot easier to preserve your good work than restore it. Copper jaws are easy to resurface when they get chewed up.

Copper jaws in the vise can be used as a heat sink for welding delicate parts (Figure 4-50). For the love of God, use copper instead of brass. Brass is hard and slippery; it makes terrible vise jaws. Copper is soft and springy—just the qualities we want in a vise jaw. Be sure to mount the vise so a long part clears the bench below the jaws (Figure 4-51).

Hacksaws. Be sure to hang a hacksaw as close to the bench vise as possible. For small cuts, a hacksaw for a quick cutoff at the bench saves time over the walk to the band saw.

For right-handed people, hacksaw on the right-hand side of the vise. It will save you many sets of barked knuckles.

Keep two or more hacksaws handy with different blade pitches (Figure 4-52). One should be 32/teeth per inch and the other as coarse as you can find, like 14/teeth. This saves changing the blades, which nobody likes to do anyway.

Figure 4–52 Hacksaws with different blade pitches.

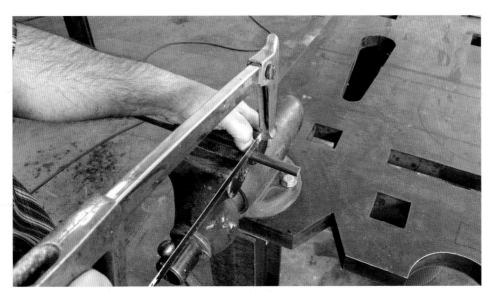

Figure 4–53 Your fingernail can serve as a guide fence.

The hacksaw will beat the cut at the power saw, including the walk to the band saw, up to .75 diameter. Most folks stand at the horizontal band saw, waiting, mesmerized by the slowly circulating blade. So save the walk and hacksaw your small items.

Start your hacksaw cut using your fingernail as a guide fence (Figure 4-53). It gets a groove started in the exact spot and keeps the blade from jumping out. Put the part you are keeping in the vise, slightly below the top of the jaws (Figure 4-54). This will keep you from spoiling the part, if you slip out of the saw groove.

Put a twang in your game and use only quality high-tension hacksaw frames. No others are even worth the energy to throw them in the scrap bin.

Figure 4–54 Hack sawing in the vice.

4.12 Filing

Many people overlook filing as an important skill; they pick up a grinder instead. Both bad filing and bad grinding can be seen from halfway across the shop. (As my old teacher would say, "A blind man would be pleased to see it." His point? A blind man would be pleased to see anything!)

Hand filing is a skill learned over a long period of time. A major advantage over powered methods is extreme control. When you cannot afford to make a mistake removing material in a delicate area, use a file. The file is directly connected to the best computer on earth. With a power tool, by the time the order to stop makes it to central command, it's often too late. Just look at the work an experienced die-maker does to see what can be done with files and hand work.

When I took metal shop in high school, we had to learn proper filing. The shop teacher was quite thrifty and would replace the files only if they made better cake frosting knives.

Whenever a new file was placed in the general population, my little trick was this: if I got the new one, I returned it to one of the extreme ends of the file rack. The next day I would swoop in and grab the end file while the others were fingering each file to find the new one they were sure was in there.

Our first filing project was a perfectly square cube. We were given a hunk of cold rolled steel that looked like somebody held a rat by the tail and gnawed it off the end of the bar. We were then required to file and sand it into a perfectly square cube of specific dimensions.

If you ever need a devilish little project to keep an apprentice out of your hair for a month or so, this is it! It sounds simple, but getting all the

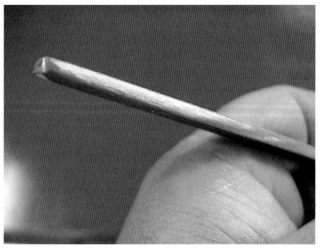

Figure 4–55 Grind a safe edge on a primary file.

sides to size and square is non-trivial. Try this with a file about as sharp as a putty knife and see what I mean.

Grind at least one safe edge on all your primary files (Figure 4-55). This reduces the chances of gouges and keeps you from scratching when filing up to a shoulder on your nice machined parts.

Round the noses of all your files (Figure 4-56), in particular any files used in the lathe. There is no reason to have the nose of the file square and sharp unless you scrape paint with it.

Figure 4–56 Round the noses of your files.

Figure 4–57 Half round files fit.

Put handles on all your files. This gives you much better leverage with the tool and makes them much more efficient. Never use files without handles in the lathe or other power machinery!

Having handles lets you hang your most-used files in a professional looking rack and keep them at your fingertips (Figure 4-58). I use these bright blue handles so that I can spot my files from across the shop.

Many people don't put handles on their files so that their files will lay flat in the drawers of the toolbox. Resist this tendency.

If you go in the field with a small toolkit, take at least one file. Make it the very versatile half round with a bastard or smooth cut. Half rounds fit into acute angle corners well (Figure 4-56) and will dress a radius as well. It's like three files rolled into one.

If you have to file a radius smaller than the file radius, use very light pressure and twist the tool as you file.

Get rid of dull files. They actually do wear out. They can be sharpened chemically if you have sentimental attachment to them.

Impress the apprentices by generously giving them your dull castoffs, When you hear them respond, "Gee, thanks mister," resist the temptation to laugh, at least until they walk off with the misty look still in their eyes.

Grind an edge on a file for knocking berries loose after welding (Figure 4-59). The finish from knocking berries off with a file looks more professional than chasing around the entire weldment with a grinder.

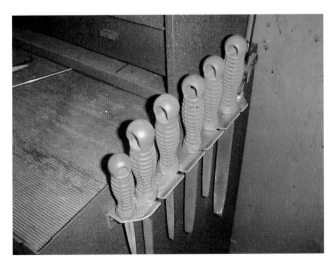

Figure 4–58 Treat your files professionally.

Figure 4–59 Your file can be used to knock weld berries loose.

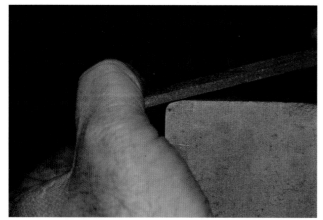

Figure 4–60 Filing a radius.

Figure 4–61 Keep the file in flat contact through the arc.

When filing a radius, follow the radius through the arc with the file kept in flat contact through the arc to keep it true (Figures 4-60 and 4-61).

The lines from filing follow the radius around the corner (Figure 4-62).

The file needs to be kept flat against the work so the corner is not rolled off to the sides.

Figure 4–62 Lines from filing.

Can I have a new file?

My old teacher Doug controlled the hand tool inventory at the shop where we worked. We did a lot of stainless, so every so often you would need a new file. Everybody knew the unlocked drawer they were kept in, yet nobody helped themselves. Somehow you knew Doug remembered the exact date you got your last file and did a lightning fast cost-value analysis on your personal file consumption. If you were trying to jump the gun or hoard a new file, you would be under close scrutiny.

The normal drill was to ask Doug for a new file when you needed one. This was no small effort. The real frustration came when he would give you the used file off his workbench when you asked for one. It was a quicker way to get rid of you and get back to work.

Everyone knew a hoard of brand new files was a mere two feet away. With those so close, who would want somebody's sullied hand-me-down file? But Doug would thrust an old file at you and say "Off you go." No discussion, end of story. I sulked back to my bench, depressed at the thought of three more months of filing deprivation.

A couple of months later I noticed another guy heading to get a replacement file. I got a brainstorm to go in right after him and get a new file. There was risk of being branded a copycat, but it was worth it. I would get a fresh one out of the drawer, knowing the handout system so well.

I had that little smirk on my face when you think you're smarter than everybody else. I gleefully watched the other guy crawl off with his used file! But, I wiped the smirk off my face and headed up to the poop deck.

"Hey Doug, can I get a new file? My old ones had it from all those stainless tanks." He looked at me with an unwavering squint of suspicion. "Didn't I just give you one a while back?" It wasn't really a question. I managed to squeak an answer, "Umm, I don't think so. It's been a few months at least."

I had carefully positioned myself between Doug and the drawer with the files. Almost there, hold on a little longer, I said to myself. "Alright, go on then, fetch one out of that drawer. Be quick about it. I don't want the whole shop beating a path to my door for files."

(continued on page 90)

(continued from page 89)

It was hard to keep the winning smirk off my face. I turned and squelched a little snicker. I opened the drawer and looked down in horror.

The box was empty.

I pawed around the inside like a dog frantically digging at a gopher hole. When I finally came up for a breath of stale air, I said,

"Hey Doug, the box is empty." I tried to hide my utter disappointment. Doug responded, "That's what I was trying to remember. Here, take this one." He handed me a sad–looking, used file from the recesses of his bench. "This one's got plenty of life left in it. Off you go now."

This is how I learned to appreciate a new file. From that day on, every time I use a file, I can't help but think of how a nice, new one cuts.

More Tools. Use a cordless drill to hand tap through holes (Figure 4-63). This is at least a thousand times faster than twirling by hand for a dozen holes. It beats setting up the tapping head for a few holes or when there are a few different sizes.

Heck, use two drills if there is more than one size hole to tap.

The block in Figures 4-63 and 4-64 is a tap guide to keep the tap perpendicular to the part. These can be made quickly for those jobs that don't lend themselves to a universal type block.

Figure 4–63 Using a cordless drill to hand tap through holes.

Figure 4–64 Using a block as a tap guide.

Figure 4–65 Indicate the age of rechargeable batteries.

Mark the age of your re-chargeable batteries (Figure 4-65).

These batteries eventually wear out. If you're like us and have a bunch of them, it's hard to keep track of which ones cannot hold a charge and should be replaced.

These can usually be taken back to a store that sells rechargeable batteries and returned for recycling. Typically they last for only 500 charges or so at best.

Small cantilever clamps make great tight quarters tap wrenches (Figure 4-66). The jaws are even grooved to hold a tap securely.

Figure 4–66 Using cantilever clamps as tap wrenches.

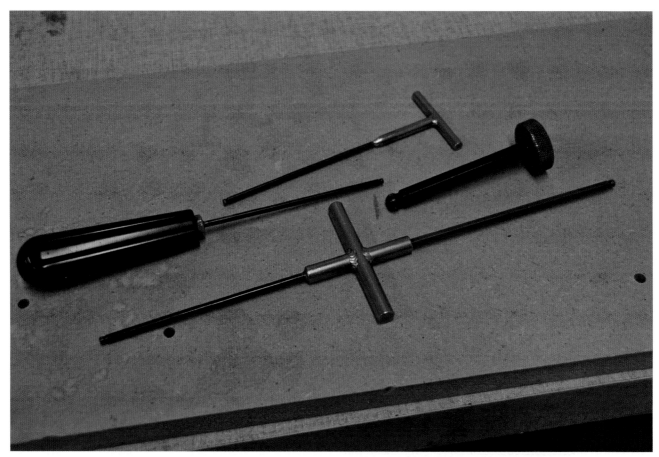

Figure 4–67 Special speed tools.

Figure 4-67 shows several special speed tools I have made over the years. We try to use as many of the same size fasteners as possible when we design machinery. I made up these ball driver tee handles for the most frequently encountered sizes. The short, fat one is for removing the socket head screws that hold the jaws in a Kurt vise. The hex blades are silver soldered into the handles and can be replaced as needed.

Figure 4-68 shows a couple of simple tap and drill extensions you can make when you have to tap a long distance. These will reach into those deep, dark recesses you have to tap a thread or just chase an existing thread. The taps are silver soldered into the extension. When they get dull, you just heat it up and pop in a fresh tap.

Be sure to make the corresponding tap drill with an extension also.

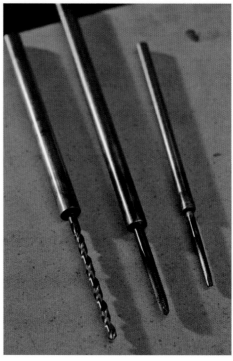

Figure 4–68 Simple brazed tap extensions.

4.13 Saws and Sawing

Every metal working shop usually has at least three kinds of saws. If they don't, their process methods should be reviewed. Saws are extremely efficient at cutting almost any material. Use the efficiency example of the metal supply center or lumber processing industry. These thin kerf methods have been optimized over a long period of time; saws are a critical part of their daily operations.

Many shops underutilize their sawing dollars by machining or grinding away material that could be more efficiently sawed away.

Don't overlook the chunk scrap produced by sawing—it has a higher per pound value than turning or milling chips.

Factors that make sawing so efficient are a thin kerf that wastes less material and uses less horsepower to remove material. Compared to milling, sawing is quite rapid. By the way, when you buy raw material from a material vendor, you pay for the saw kerf that goes in their scrap bin, but you pay good material price for it. These guys have had years to figure out all the kerfs and profit angles.

Vertical band saw. If you're just setting up a shop and are deciding what saw to get first, I strongly suggest a vertical band saw. These highly versatile saws can cut almost anything. With a few different blades, you can cut anything—meat, hardened steel, even glass. Key factors in selecting a vertical saw are speed range and throat depth. Buy the maximum you can afford in both factors.

A blade welder is nice, but not a deciding factor. Commercially welded blades are cheap in comparison to welding them in the shop.

It's nice to cut a blade and insert it into an internal cut, so the ability to connect a blade quickly is an excellent skill to develop.

TIG welding. If you don't have a blade welder or you ran out of silver solder, you can TIG weld band saw blades with silicon bronze rod (Figure 4-69). The weld is annealed a second time after grinding to remove brittleness along the heat affected zone (HAZ). Accurately line up the spine of the blade or it will make goofy clicking noises as it runs through the guides. This example shows a special right-hand left-hand blade (Figure 4-70).

Figure 4–69 TIG welding band saw blades with silicon bronze rod.

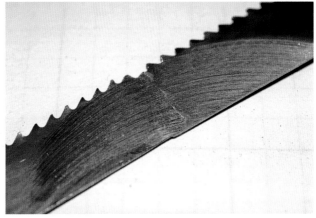

Figure 4–70 A right-hand left-hand blade.

Figure 4–71 Exposing the weld area.

Figure 4–72 Using an abrasion device to remove buildup and level the seam.

After you weld, silver solder, or otherwise connect the two stray ends of a band saw blade, remove the weld buildup and level the seam. Use a curved surface to expose the weld area to your preferred abrasion device (Figures 4-71 and 4-72).

After grinding, peen the weld area on a flat dolly to level the seam so the blade thickness is correct (Figure 4-73).

Figure 4–73 Peening the weld area.

Figure 4–74 Protecting surfaces with tape.

Figure 4–75 Surface Armor protects important surfaces.

Protecting surfaces. Protect critical surfaces with tape before cutting (Figure 4-74). A good grade of masking tape also works well as a cutting guide. The clear sheet shown in Figure 4-75 is called Surface Armor and is available in many different tack strengths. This material protects your important surfaces. It also removes without leaving behind a residue.

Lay a couple of lanes of protective tape on the saw table to keep your nice material blemish free (Figures 4-76 and 4-77).

Figure 4–76 Use several lanes of protective tape.

Figure 4–77 The tape protects your material from blemishes.

Figure 4–78 Paste a full-scale drawing to the part.

Full-Scale Drawings. Paste a full-scale drawing to the part and band saw without layout (Figures 4-78 and 4-79). Use a little spray contact cement to attach a drawing directly to the part. This is a great trick for prototyping. Laser printers are extremely accurate. I use prints from them for radius gages and checking templates all the time.

Figure 4–79 Band saw the part.

Figure 4–80 A wood
cutting circular saw.

Circular Saws. Use a wood cutting circular saw to cut aluminum plate (Figures 4-80 and 4-81). Be sure to wax the blade.

Use the cheapest carbide-toothed blades you can find at your local home center. High-quality blades don't seem to hold up any longer than the cheap ones. I have personally cut 2-inch thick copper with this setup. If you use a guide, you almost have a plate saw.

Figure 4–81 Using the saw to cut aluminum plate.

Figure 4–82 A wood cutting miter saw.

Miter Saws. A wood cutting miter saw makes short work of aluminum and plastics (Figure 4-82).

We added a fence with a recessed scale to keep from scraping the stick-on ruler off (Figure 4-83). Four-inch diameter aluminum is a snap with this setup.

Use a carbide-toothed blade for cutting aluminum. Wax the blade carefully for best results. For all the blades we tried, there was little difference between the expensive high end blades and the rough construction grade blades.

Figure 4–83 Adding a fence with a recessed scale.

Saw Blades. One of the great abuses of band saws is not changing the blade for cutting thicker or different alloys of material. If I had unlimited floor space and a budget to match, I would have three vertical band-saws, each set up differently. But I don't live on Fantasy Island. Therefore, I have to change the blades along with everybody else. I don't know why people don't get this one. The chart has been on the front of the machine as long as I can remember, but here it is, condensed to:

Thin stuff = Lots of smaller teeth   and   Thick stuff = Fewer but bigger teeth

To better understand, look at the blades they use to saw trees into lumber. These have about a 4-inch pitch between each tooth; the gullet is big enough to park a chip the size of a cinnamon roll. For most practical purposes, the average shop can get along with two blade pitches. Having these two will cut down on the confusion for people who don't look at the blade chart.

Friction Cutting. Use your old wasted blades for friction cutting sheet metal (Figure 4-84). Runing the machine at the highest possible speed (and wearing ear protection) will make short work of cutting sheet metal of every flavor up to about 10-gauge for average horsepower machines.

This approach also works for cutting hard stuff like linear bearing shafting. The process is not like sawing because the material is heated and softened to its melting point and carried away by the moving blade.

Figure 4–84 Friction cutting with old wasted blades.

Figure 4–85 A heat discolored burr.

In Figure 4-85 you can see the heat discolored burr in 10-gauge stainless steel. The process is limited by available horsepower when you get into the thicker materials. The main trick is using a high SFM cutting speed, so run it as fast as your saw will permits.

Figure 4–86 Cutting square corners in a tight spot.

Figure 4–87 Cutting the straight sections.

Cutting Square Corners. Figures 4-86 through 4-90 show a trick for cutting square corners in a tight spot.

Blades that are 1/2 wide are a good choice and cut very straight in the straight sections (Figure 4-87), hold up for long periods of time, and are not as sensitive to blade tension issues as narrow blades. If you want to make a good compromise and pick one size, then make it 1/2. The tradeoff is less turning ability. Band saws are superb roughing tools.

Start your cut and just steer for the opposite corner (Figure 4-88). Come up right into the corner to establish its position; then back out.

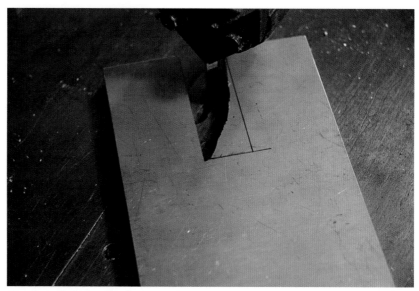

Figure 4–88 Steer for the opposite corner.

Figure 4–89 Repeat for the opposite corner.

Figure 4–90 A good band saw works well.

Do the same for the opposite corner (Figure 4-89).

Alternate this way until you can start the blade in the short direction. Use the front of the blade and work sideways like a die filer to clean out small features. Use a good fresh blade for best results. Who needs a milling machine anyway when you have a good band saw (Figure 4-90)?

Block-Shapes and Rounds. Cut large block-shaped parts out of large rounds (Figure 4-91). If large blocks are not readily available, rough saw them out of large rounds, which are sometimes the only form you can get on short notice.

Figure 4–91 Use large rounds to cut large block-shaped parts.

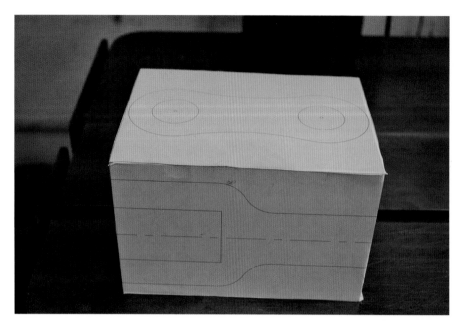

Figure 4–92 Bond laser printed drawing to the raw material.

Three Dimensional Sawing. Three dimensional sawing can save much time in the machine shop. Leave enough stock for cleanup cuts when you use this method. Better to take extra passes than have a spot with raw saw marks. In Figure 4-92, I have laser printed the drawing full scale and bonded it to the raw material with spray contact cement.

Holes and corner reliefs are added next (Figure 4-93). First cut features that allow you to keep the blank as block-like as possible (Figures 4-94 and 4-95). The danger is that you cut away your guide lines or end up with something difficult to manipulate in the saw. The vertical band saw is a very efficient metal removal tool.

Figure 4–93 Add holes and corner reliefs.

Figure 4–94 Cut features that keep the blank block-like.

Figure 4–95 Continue shaping the part.

Making a Belly Board. Forget one of those wimpy pathetic television machines for working out your abdominal muscles. Make a belly board and cut some 1-inch stainless steel for a few hours (Figures 4-96 and 4-97). A belly board is a short length of scrap wood cut to fit between the saw operator's torso and the work being pushed through the vertical bandsaw. The length is dependent on the size of the part to be sawed and the size of the operator. The idea is that your body weight is pressed against the board to provide the large forces necessary to cut difficult materials like stain-

Figure 4–96 Making a belly board!

less steel, while leaving your hands free for the fine motor control needed to accurately steer the cut. The board should be wide enough to spread the force enough so that heavy pressure can be used and still be comfortable. Six pack abs here I come, or maybe it's just a six pack here I come...

Figure 4–97 Cutting 1-inch stainless steel.

Figure 4–98 Sweep from behind the blade toward yourself.

Using a Push Stick A great practice to get used to is using a push stick to sweep the scraps off the saw table. It is important that you develop the habit of sweeping from behind the blade toward yourself and the operator side of the saw (Figure 4-98). If you were to accidentally hit the blade, the last time I checked there were no teeth on the back of the blade. You will get a good scare, but still have all your fingers intact. Check out the picture of the push stick in Figure 4-99. I wonder how it got so many saw cuts in it. You think it is an effective tool?

Figure 4–99 A push stick protects fingers.

Figure 4–100 A vertical band saw vise.

Band Saw Vise Figure 4-100 shows a great vertical band-saw vise we found recently. You can clamp those pesky jobs where you are more worried about clipping a finger than doing a good sawing job. Short pieces of round bars are easily clamped for a spin-less, stress-free cutoff (Figure 4-101).

Figure 4–101 Clamp short pieces of round bars.

Figure 4–102 Grinding the weld edge preparation.

Preparing to Weld. Grind the weld edge preparation with the teeth facing opposite directions (Figure 4-102). Twist the blade 180 degrees, then grind the ends square, prior to welding. The grind angles will match perfectly for a perfect weld (Figure 4-103). After grinding, I like to add another quick anneal to the weld area to remove brittleness that occurred during the weld finishing (Figure 4-104).

By the way, I used an old friction blade for the photo. We don't weld many blades in house anymore, and let somebody else do it.

Figure 4–103 Preparing for a perfect weld.

Figure 4–104 Adding a quick anneal to remove brittleness.

Figure 4–105 Cutting small rods in the horizontal saw.

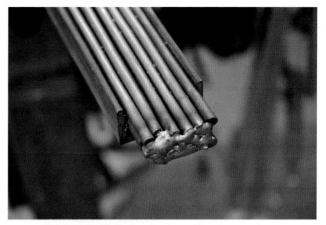

Figure 4–106 Weld the ends of small rods together.

Figure 4–107 Preparing for smaller cutoffs.

Cutting Small Rods. If you have a bunch of small rods to cut in the horizontal saw, drop them in a channel and weld the ends together (Figures 4-105 and 4-106). This keeps the rods from rotating or changing lengths while you do multiple cutoffs (Figure 4-107). By grouping small sizes together, you can now take advantage of the free cutting abilities of coarser pitch blades to make short work of a large number of pieces (Figure 4-108).

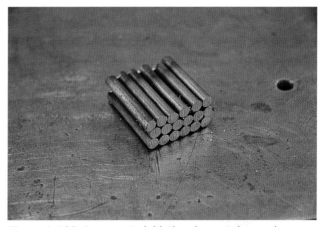

Figure 4–108 Coarser pitch blades shorten the work.

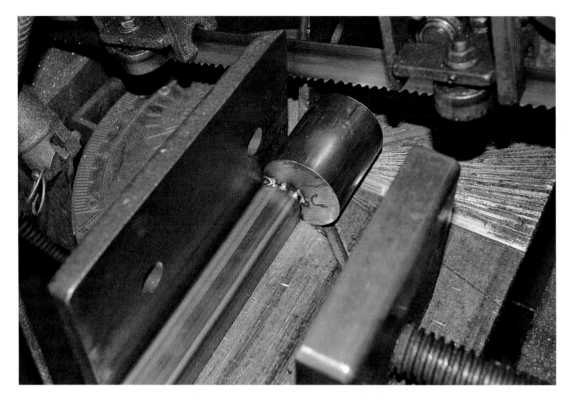

Figure 4–109
Weld a short
stump to
another piece.

Welding Short Stumps. Weld your short stump of material to another piece to get another couple of cuts in the saw (Figures 4-109 and 4-110). Or you can use a jaw spacer to keep the vise jaw from tilting (Figure 4-111).

Figure 4–110
Making additional cuts in short stumps.

Figure 4–111
Using a jaw
spacer.

When making multiple cuts on angle, stack the pieces so the vise pressure clamps the lengths together (Figure 4-112). Three-point contact makes for firm clamping.

Figure 4–112
Preparing for
multiple cuts
on angle.

Figure 4–113
Cutting miters
in the saw.

Cutting Miters. When you cut miters for weld prep, cut the angle just a fraction of a degree more or less than 45 (Figures 4-113 and 4-114). This ensures the joint is in contact either inside or outside.

Figure 4–114
Use a degree
more or less
than 45.

Figure 4–115
Tack welding
corners that
are in contact.

When you tack weld corners in contact (Figure 4-115), you can move the frame a bit to square it accurately (Figure 4-116). If you cut parts so the inside of the frame touches first, you save joint preparation time on the outside. Note which side you tack first. You want the tack to act like a little hinge when squaring.

Figure 4–116
Squaring
a corner
accurately.

Figure 4–117
Cutting a cross
member a
little short.

Cross Members. Sometimes it's easier to cut a cross member a little short instead of grinding a big weld preparation (Figure 4-117). This also pays dividends when finish grinding because the finished weld is closer to flush with the surface (Figure 4-118).

Figure 4–118
Finish grind-
ing the cross
member.

Figure 4–119
Using two
layout lines
with a vertical
band saw.

Layout Lines. When you need to do a really accurate job on the vertical band saw, try using two layout lines with a spacing as close as possible to the blade size (Figures 4-119 and 4-120).

When you follow a single scribe line, it's easy to wander double the width of the blade and still stay on a single scribe line. It's like driving a car between two telephone poles and not clipping your mirrors. This trick works well for splitting a two-piece part for which the designer has not left enough sawing and cleanup allowance.

Figure 4–120
The spacing
is close to the
blade size.

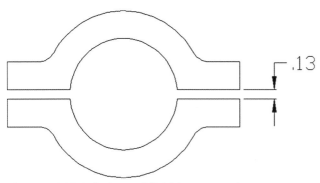

Figure 4–121 Splitting and finishing a two-piece part.

Figure 4–122 Creating a split part with a zero gap.

Split Parts. A minimum starting number for splitting and finishing a two-piece part should be 1/8 for easy materials and moderate thickness up to 1.00 or so (Figure 4-121). The thicker the stock, the wider the split needs to be. Two methods can split parts quickly and cleanly in the average shop. In most cases, you can use a slitting saw and slice right through without the need to re-surface the cut. You could also band saw and re-surface the cut face. The problems come up when there is too little allowance for cleanup.

Thicker slitting saws are more stable in deep cuts and can be fed more efficiently than skinny wobbly large diameter saws.

One trick for creating a split part with a zero gap out of one piece of material is to start out with a diameter slightly larger than the finish size, and split in half first (Figures 4-122 and 4-123). In many cases, the designer needs a part that has the split exactly on the centerline, which makes for twice the work and twice the material in the shop. With expensive material or large sizes, this can cause a big cost increase.

Figure 4–123 Start with a diameter slightly larger than the finish size.

Pre-splitting works well if the part has features like screws or pins to locate the two halves together for the finish tuning and boring, like the Tantalum part seen in Figure 4-124. A bar a bit larger than the finish diameter was split and then held together to create this centerline split. In this case, wire EDM was used to split the part to minimize material loss. In some cases, parts can be tack welded together if they don't have any fasteners; the last operation removes the small welds. This saves making two parts to get a complete assembly with an exact centerline split.

We have discussed several ways to get your shop up and running efficiently. Don't underestimate the importance of the basic setup and infrastructure in your own shop. To reduce overhead and increase profit and efficiency, you have to get into the small details of how things get done out on the shop floor (Figure 4-125). Good leaders and owners should question their methods from time to time and ask the question, "Is there a better way?"

Figure 4–124 Using screws and pins to fit two halves together.

Figure 4–125 Don't put the book down now!!!

4.14 Rigging and Lifting

"Humans were invented by water to carry it uphill."

This section deals with lifting and moving things too heavy to move or carry manually. From time to time, you will be surrounded by heavy machinery and materials that need to be moved. This can be tricky and dangerous work for the inexperienced. The ability to handle and secure loads safely and quickly is an integral part of the job process; you may need to handle these items multiple times during construction. The more efficient your handling, the faster the job will completed.

This subject is sometimes called *rigging*. In the old days, riggers were the guys on the ground securing slings and loads for the crane operator. In the small shop or home garage, the rigger might be a professional hired to move that milling machine you just bought—or it could be you working alone, trying to get that same machine home on your rented trailer.

Whatever your situation turns out to be, slowly and with careful consideration of the consequences is required. My best advice for the small shop and hobby machinery movers is to go really slow. Think pyramids when you have to move a machine. Most of the mishaps occur when trying to go too fast or take a shortcut. The internet is filled with hours of this kind of entertainment.

Humans are basically pretty weak animals. Without our tools and machines, we are just hairless primates with no particular strengths other than our big brains. Rigging and lifting in the shop requires brains. In large shops, cranes, hoists and forklifts seem like pretty simple machines that anybody should be able to use without any difficulty. This is pretty far from the truth.

In my years of experience, it seems the simplest machines and processes cause the most trouble. The ease that you move objects that weighs several tons, with no sensation of weight or effort, disguises the seriousness of these operations. They are taken for granted and that is when trouble starts.

If you're lucky enough to have some kind of hoist, forklift, or other material-handling equipment, be sure that everyone who uses it has been checked out in its use or had instruction and supervision from someone more experienced. There are many subtleties and serious consequences to moving heavy things with small brains. The saying goes, "The rigger must not guess, he must know."

Before you start, know the weight, the equipment capacity, how to retain the load, and how you're going to control the load. These are the key elements for success in moving heavy objects.

"Bend your knees when you lift something heavy." The same consideration should be given when preparing to move something too heavy to lift personally. Think before you lift; protect yourself at all times. Moving heavy things is like getting into a cage to feed a hungry tiger. Move slowly, have an escape route, and trust nothing.

Whenever we handle, move, or load equipment or materials, it is the responsibility of the rigger to make sure the entire process is carried out in a safe manner for the next person in the job process. This includes landing the load so it is safe, as well as securing the load for safe transport or, at a minimum, supervising the work to ensure proper execution by others. Typically the rigger has overall responsibility during a lift or move. It's important to note that only one person should give instructions where multiple people are involved.

Figure 4-126 The bowline.

Figure 4-127 The trucker's hitch.

Rope work. Everybody should know how to tie at least two knots. My two favorites that can be used almost anywhere for multiple purposes are the good old bowline (Figure 4-126) and a simplified version of the trucker's hitch (Figure 4-127). The first problem that seem to plague people when they pick up a length of line is, "I need a loop." The second problem is, "I need to get the slack out of this line and get it tighter than all get out." Everything else falls in line behind these two needs. There are hundreds of useful knots, but what I tell anybody who asks me is just learn two. If you remember and master these two, then you most likely won't need any more.

The bowline is pretty simple and a great way to get a loop in the end of a line. They will untie easily even after subjecting them to heavy loads. I used to sail on a large sailboat and the jib sheets where attached to the clew of the jib with a simple bowline. The sheet was 5/8 diameter Dacron line. Keep in mind that this is how the wind connects to the boat and drives it through the water. I used to stand on these sheets sometime just for fun. I weigh over 200 lbs and my weight would barely deflect the line; it had so much tension in it. It was more like a steel rod than a piece of rope. After docking the boat, the jib sheet bowlines could be easily untied in two or three seconds even after that severe loading. This is the champion of knots.

The bowline is a simple enough knot that there should be no excuses why everybody can't tie one. It's the old story about the rabbit coming out of the hole (Figure 4-128) and going around the tree and back into his hole (Figure 4-129).

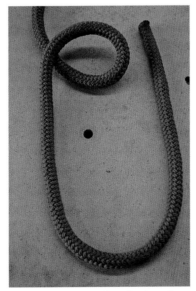

Figure 4-128 Starting the bowline.

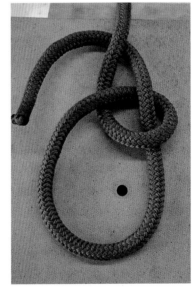

Figure 4-129 Completing the bowline.

Figure 4-130 Starting the trucker's hitch.

Figure 4-131 Cross the free end over.

The second most useful knot that I have come to use over and over again is the trucker's hitch. There are several variations of this knot that work but are not easily untied or changed quickly. The beauty of this particular method is you tie no semi-permanent loops in the line. The loop is secured by a half hitch and just falls apart when you untie the load. Start out by passing the line from the load and securing it around the hook or load eye (Figure 4-130). I'm right handed so I go from left to right as shown.

Cross the free end over, as shown in (Figure 4-131) while holding the bit from the load. Pull up some slack from the free end to form a loop (Figure 4-132). This will become your "pul-

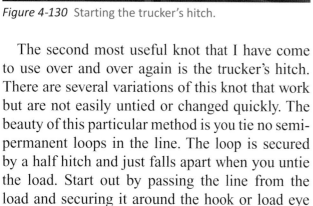

Figure 4-132 Form a loop.

ley" which will tension the load line. Twist the load line toward yourself and form an eye (Figure 4-133). The loop in my left hand (shown in Figure 4-132) is inserted into this eye (Figure 4-134).

Figure 4-133
Form an eye.

Figure 4-134 Insert the loop into the eye.

Figure 4-135 The simplified trucker's hitch.

Figure 4-136 Forming the pulley.

Figure 4-137 Place wood before moving a heavy object.

Figure 4-135 shows the simplified trucker's hitch tied properly. As you pull on the free end, the half hitch eye tightens around the loop and forms the pulley used to tension the load (Figure 4-136). Once the load is tightened, the free end can be secured around the fixed end loop with a couple of half hitches. If you really need more tension, you can double this trucker's hitch up to get twice the tensioning power. The double trucker's gets a little messy for concise pictures, but try it once you have the single trucker's hitch down pat.

More lifting. Whenever you have to move a piece of material or heavy machinery with the forklift or anything that has a steel or metal bottom, always put a piece of thin wood between the metal surface and the object that is being moved. (Figure 4-137). All you need is a thin piece of plywood which acts like a brake shoe and keeps the load in position. The oily bottoms of machine tools are a good example of metal-on-metal as a low friction bearing surface. A metal-on-metal moving scenario can mean the difference between the load shifting dangerously and a simple lift. I'm sure readers who have used forklifts have experienced a long piece of bar stock pivoting as if on bearings on the fork blades as a small turn was made. Uncontrolled loads are just that—uncontrolled.

Figure 4-138 shows a fast simple easy to make a lifting lug without a hole. It's formed from stock flat bar sizes with two equal angle bends. Please note this type of lifting lug is not suitable for turning loads over; it is used for pure vertical lifting like you would find in tanks and boxes. It is super easy to make and add on where you need it during fabrication (Figure 4-139). These lugs work equally well with wire or rope slings (Figure 4-140). For loads that need to be turned over, you will need a weld on a lifting eye with a closed hole.

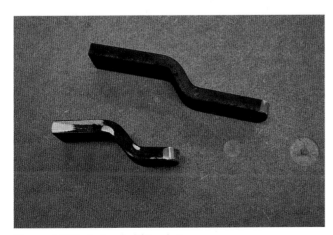

Figure 4-138 A simple lifting lug.

Figure 4-139 Add to the lug as needed.

Figure 4-140 Works with wire or rope slings.

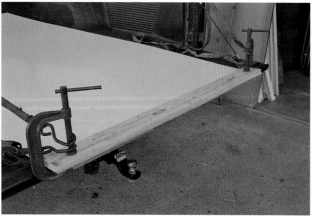

Figure 4-141 Securing sheet metal with clamps.

Figure 4-142 Use wood backing for thinner material.

How to secure sheetmetal or other flat sheet material in a pickup truck. The best method of securing sheet material in the truck is to strap it to its shipping pallet or skid. Many times this is not possible and individual sheets or plates need to be transported. One trick we use is to secure the sheet with clamps, then use the clamps as tie points for the load (Figure 4-141). For thinner material that might be damaged by the clamps, a wood backing board can be clamped to the sheet for greater security (Figure 4-142).

Three-point leveling and moving (Figure 4-143) is superior to four-point leveling in every way. It's easier and faster to level something with three contact points compared to four. Three points are like a stable tripod and are always in contact with the floor or table no matter how uneven it is. Four points are subject to tilting or rocking, which, when leveling, is confusing. Moving a heavy load or working with a rough floor can make a simple leveling or moving job a real headache, trying to balance on four points. In Figure 4-144, I am using two fingers to move a heavy table set on three points.

Figure 4-143 Three-point leveling and moving.

Figure 4-144 Easily moving a heavy table set.

Figure 4-145 Skateboard dollies.

Figure 4-146 Turning and steering the dollies.

Skateboard dollies. I call the devices in Figure 4-143 skateboard dollies. They're made from common radial ball bearings and a little steel in a couple of hours. They make short work of moving heavy objects. Years ago I saw a millwright use a similar set to move several machine tools out of a tight spot; I had to have my own set. Used on relatively smooth concrete floors, one person can move several tons of machinery with a lightweight pry bar or a gentle push.

Slip a piece of wood between your machinery and the dolly. If you use three dollies under your load, they will always stay in place. When using four dollies, check their placement as you move the load. Uneven floors and four contact points can cause tilting where the weight is completely removed from one corner, allowing the dolly to slip out or under the load. The round bar sticking out in Figure 4-146 is to turn and steer the dollies with a length of pipe.

Figure 4-147 Various wedges.

Wedges. Wedges have the ability to slip into places other kinds of tools won't. Figure 4-147 shows several types and sizes that are handy around any metalworking shop. The humble wedge can generate forces beyond belief when driven with a hammer. The thin tip of a pipe flange wedge can open enough of a gap that a pry bar can then be inserted (Figure 4-148). You can also stack steel setup wedges back to back to make quick height adjustments or create a parallel lift or wedging action (Figure 4-149).

Figure 4-148 A pipe flange wedge.

Figure 4-149 Stacking steel setup wedges.

Figure 4-150 Generating force with wedges.

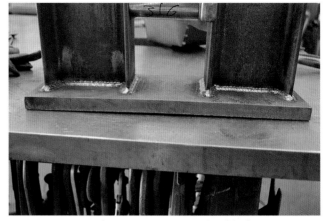

Figure 4-151 Wedges generate considerable force.

Figure 4-150 and 4-151 illustrate just how much force can be generated with wedges. Two steel wedges back to back and a few taps with a hammer easily bow the thick steel base plate. Wedges can be used to create gaps or openings or keep them from closing down during welding.

The compact hydraulic rams seen in Figure 4-152 pack a lot of power in a small package. Extensions and a variety of end fittings make these useful in the fabrication shop. With 10,000 psi hand pumps, huge forces can be generated without your realizing it. A useful addition for straightening work is a pressure gauge so you can have some feedback as to how much force you are applying. Always use extensions that are positively secured into the unit. This is another place where a thin piece of wood can keep the oily base from sliding out of position unexpectedly (Figure 4-153). I once saw somebody's front teeth get knocked out from a chunk of wood they were using as a makeshift extension. He was asked to leave the woodchuck club….

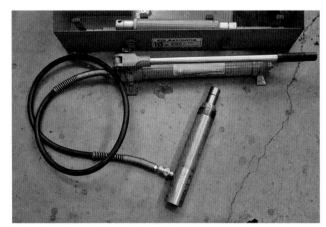

Figure 4-152 Compact hydraulic rams.

Figure 4-153 Wood keeps the oily base from sliding.

Figure 4-154 Basic pry bars.

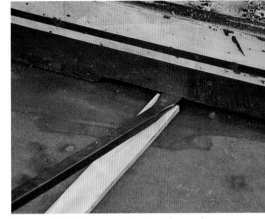

Figure 4-155 Keeping the bar from slipping.

Prying and levers. Figure 4-154 shows several types of basic pry bars. You will need a couple of different tip configurations for the typical work in welding and machine shops. A little block of wood or sliver of plywood keeps the bar from slipping during a heavy pry (Figure 4-155). My all-time favorites are the screwdriver handle type. I seem to find a million uses for these almost unbreakable prying tools.

Here is a method for weighing a large object that won't fit on the scale or is outside the range of your weighing device. Hopefully you have a relatively level area where you can do this. Simply weigh one end at a time (Figures 4-156 and 4-157). Then add the two weights to find the total. Be sure the opposite end is the only thing touching the ground. If your object must be kept level you can put a block the same height as the scale under the opposite end. The sum of the weights of the ends is the total weight of the object. Don't believe me? Stand on two scales with one foot on each scale and add the two weights.

I hope you found a few tidbits in this section on rigging and lifting that might save you on some heavy lifting. Remember: go slow and be safe when handling loads heavier than you would care to drop on your foot. This is one skill you don't want to learn the hard way.

Figure 4-156 Weigh one end of a heavy object.

Figure 4-157 Then weigh the other end.

LATHE COMPOUND

$84.3° = 10$ to 1 REDUCTION

$63.4° = 2$ to 1 REDUCTION

(ARC-TANGENT) $10 = 84.3$

Manual Lathe

5.1 Learning to Love the Lathe
5.2 Getting Started with the Manual Lathe
5.3 Step Turning
5.4 Threading in the Manual Lathe
5.5 Multiple Start Threads

The manual lathe is the cornerstone of any machine shop. Almost all workers starting out in the machine shop finds themselves on the manual lathe. Lathes come in all sizes and shapes, as you can see in Figure 5-1. The lathe has been called the king of all machines for good reason. You can bet if you need a forklift and a ladder to put the tool bit in a machine like the big old Niles machine in Figure 5-1a, some fun is bound to happen.

Anybody who has spent time on a modern lathe would immediately recognize all the design features of Henry Maudslay's revolutionary screw-cutting lathe, which he built around 1800. It is one of the oldest machine tools and its look and features have not changed much since its invention. Another famous Englishman Joseph Whitworth added the compound rest feature which transformed Maudslay's design greatly. This was a major design breakthrough for what we now recognize as the modern engine lathe.

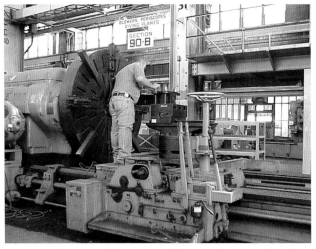

Figure 5–1a Large Niles Engine lathe in the Mare Island Naval Shipyard Machine Shop.

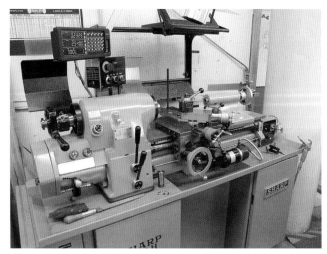

Figure 5–1b Good-sized, accurate lathe for hobby or prototype shop, with digital readout.

5.1 Learning to Love the Lathe

I started on the lathe in high school. At the time I was disappointed; I had wanted to be assigned to the welding section because I had been welding for a quite some time and was eager to demonstrate my skills.

The high school shop had four little Rockwell 9- or 10-inch manual lathes set off to one side of the shop. These lathes had the old rocker style tool posts and quick change threading gearboxes and taper attachments that took a little head work to figure out.

The funny part is that somehow the entire semester I spent on the lathes. I learned a tremendous amount and ended up enjoying the work immensely.

Our first project was a stylish aluminum meat tenderizer, a definite must have for every modern kitchen or crime scene. We got to straight turn, taper turn, thread, and knurl on this one project. Like almost all high school machines, the corners of the compound rest were hammered to death from running them into the spinning chuck jaws. Later in life I saw a great solution to this problem.

The shop teacher fabricated aluminum blocks that were glued or screwed to the compound and served as sacrificial beating blocks for the lathe newbies. This improvement increases the resale value of any lathe. Among the first things used machinery shoppers look at are the condition of the ways and the corner of the compound. A clean crisp corner on the compound is usually an indication of a gentle life.

A few years later I got my first lathe, an old 1915 Prentice with a 9-inch swing. It had a flat leather belt that made a unique tick-tick-tick sound as the joint in the flat leather belt passed over the sheaves. It came to me with a huge stack of change gears for threading. If you want to learn to run a lathe, start on something old and loose. When you can get things done on an old worn machine, you will be a superstar astronaut on a tight modern machine. It took me a week to figure out how the clever little planetary back gear setup worked. The guy I was working for "offered" to let me store it for him while he was going through a divorce. I kept it a couple of years and ended up buying it from him for $300. This was a huge sum because I was only making $3 an hour part time.

5.2 Getting Started with the Manual Lathe

Small work can be done on large lathes but large work cannot be done on small lathes. Buy your lathe a little bigger than you think you might need.

Put target score marks on raw material for fast roughing (Figure 5-2). Touch the tool and use your scale to put reference marks so you can save measuring time. Rip and tear down close to the lines, then pull out your finer measuring tools (Figure 5-3).

Use quick change type tool-posts. Be sure you have plenty of tool blocks. I hate when a lack of tooling interferes with my productivity. With metallic sharpie markers, you can write your target numbers on the tool holder instead of a paper cheat sheet.

Always leave a boring bar set up in a tool block. 1/2″ or 5/8″ are good starting points for general purpose work. Use inserted tooling if many people use the machine.

Figure 5–2 Target score marks.

Figure 5–3 Rough cut near the lines.

Figure 5–4 CCMT and WNMG inserts.

Figure 5–5 Double-ended, 45-degree chamfer bit.

Always have an inserted turning tool set up in a tool block for general purpose work if different people use the machine. CCMT or WNMG inserts are a good compromise (Figure 5-4). You can switch insert geometry easily for different materials. WNMG (left insert of Figure 5-4) gives you six cutting edges per insert for good economy. We set ours up with the thought of one optimized for harder materials and one for softer materials.

Try to set up your dedicated turning tool block so it misses the quill of the tailstock when the tool point is on center. This saves having to reposition the tool when using the tailstock to support your work. Keep two parting blades set up: one neutral and one with a couple of degrees left hand angle. Always keep a long, double-ended 45-degree chamfer bit set up at each lathe. It is extremely handy for edge breaks and quick facing (Figure 5-5). The double end allows you to use it on both axes ID and OD.

Lathe tooling. Never modify someone else's hand ground tool left in a tool-block. You might as well ask to borrow their toothbrush. You just don't do it. If somebody leaves one in a tool block, take it out and leave it on the top of the lathe.

Always leave the lathe in better shape than when you found it. This has the added benefit of high-lighting the shop slobs. They stand out in stark relief against a clean background, where they can be properly whipped and chastised.

Use a depth of cut (DOC) that is a little larger than the tool nose radius. With inserted tools, the chip breakers don't function unless the depth of cut is larger than the nose radius.

To minimize chatter, use positive tool geometry with small nose radii and lead angle near ninety degrees. This is especially true for ID boring. The tool tip should be on center or a few thousandths high.

Figure 5–8
Heavy metal boring bars.

Step bore deep bores in two or more steps (Figure 5-6). This leaves more room for chip evacuation and you can use a boring bar that fills the bore more completely (Figure 5-7). This is a very useful trick in the CNC lathe.

Carbide or heavy metal boring bar shanks are more rigid for those deep holes. You can buy the heavy metal material and make your own boring custom bars. The long one in Figure 5-8 uses old broken 1/4-in carbide end mills for tool bits. Every shop has an endless supply of broken 1/4 tools that can be re-used in this boring bar. The longer bar in Figure 5-8 is a rigid Tungsten-heavy metal alloy called "No-Chat." It can be machined into any configuration you can imagine and lives up to its reduced chatter advertising.

When you have a chatter problem, try increasing the feed rate before you try slowing everything down. Another trick is to move the boring bar in the holder a fraction of an inch in either direction. Sometimes this small change in the resonant frequency can reduce or even eliminate chatter.

Carbide inserts have a calculated design life of twenty minutes of cutting time. This is great if you sell inserts and not so great if you have to write the checks for them.

Always try to increase cutting speeds and feed rates. If you never push the envelope, how do you know where the limits are?

A 20% increase in feed rates returns a greater reduction in part cost than a 50% increase in tool life.

Test and try new tools once in a while. Lots of smart people are working on some really good tools. Besides, it's fun to test the salesman's tools at full throttle.

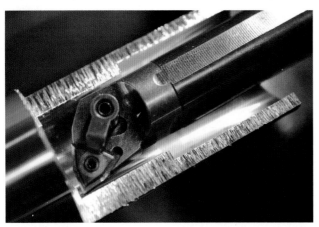

Figure 5–6 Step boring.

Figure 5–7 Using a boring bar.

Figure 5–9 Orient rough cut blanks.

Figure 5–10 The amount of taper produced varies.

Be aware of the reduction in clamping force when spinning a chuck at high speed. You can lose 50% of your clamping force at high RPM.

Orient rough cut blanks from oxy-fuel, waterjet, and plasma cutting with the large end of the taper in the cut to the back of the chuck (Figure 5-9). All of these processes produce varying amounts of taper, depending on the material thickness (Figure 5-10).

About the only thing good about interrupted cutting is that the chips break no matter what.

When roughing, try to get completely under the bark or outer skin of a bar on the first pass. This will be repaid in tool life.

Most lathes have .001-diameter divisions on the cross slide dial. A decent operator can control diameters to .0003 by interpolating between the divisions — if the lathe is in good shape.

Your lathe should repeat to half the smallest division on the cross slide dial.

Set your compound rest at a small angle to dial in increments smaller than the cross feed dial dimensions. A 5.73 degree angle off the Z axis will take off .0002 on the diameter on the X axis for each .001 dialed on the compound. When I have used this trick, I set it at an even 5 degrees to keep me on the MMC side. This is an old trick that is seldom used in the job shop, but I still like the idea.

Try to grind your special lathe tools so the tool post can stay square to the machine axes. This saves setup time, but is not always possible.

Lathe workholding. Hardinge makes excellent internal expanding collets (Figures 5-11 and 5-12). These are great for holding bushings and sleeves on the inside diameter. One inner mandrel is used for a large range of expanding collet diameters. They are still soft enough that you can turn the OD a little for one of those oddball emergency jobs that always seem to come up.

Figure 5–11 An internal expanding collet.

Figure 5–12 Internal expanding collets hold bushings and sleeves.

Figure 5-13 Drop plastic parts or bushings in hot water.

Figure 5-14 Slip parts or bushings on a gage or dowel pin.

Try this trick for holding on the ID of plastic parts or bushings. You can expand them a few tenths by dropping them into hot water (Figure 5-13), then slipping them on to a gage or dowel pin (Figure 5-14). You can then turn a precision OD by holding the pin in a collet or chuck. To remove the part, just dip the pin and bushing back in the hot water. By exploiting the fact that the plastic expands much more than the steel and faster, you can slip it off the pin.

Grind a couple of multipurpose tool bits (Figure 5-15). This can save time on manual tool changes. The one in Figure 5-15 turns ODs, faces, chamfers OD, and ID — all in one tool.

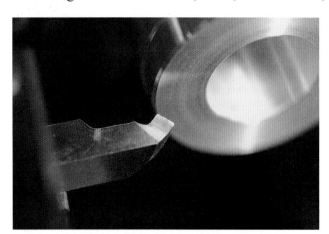

Figure 5-15 Grind multipurpose tool bits.

Make yourself an old school spring-type tool-holder (Figure 5-16) for large radii or use form tools (Figure 5-17) with a broad cutting edge, or any tool that you are having chatter problems with. The spring-type tool-holder with its pivot above the centerline backs the cutting edge off when the tool bites and starts to chatter. Typically you can increase your cutting speed by double, using a setup like this. Contrary to the normally accepted thinking, sometimes more rigidity is not the answer.

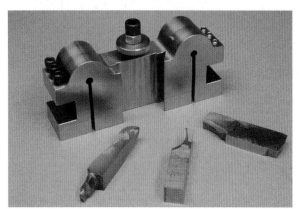

Figure 5–16 A spring-type tool holder.

Figure 5–17 Use form tools.

Figure 5–18 A very wide form tool.

Figure 5–19 A curved form tool.

In the old days, when planers were used to surface plates and other large flat surfaces, the final finishing tool was a very wide flat tool held in a gooseneck or spring-type tool holder similar to the one in Figure 5-18. Form tools are an excellent fast way to duplicate complicated geometry in the manual lathe.

Figure 5-19 shows a curved form tool that on a good day has chatter written all over it. In this example, this tool was plunged into 4140 steel (Figures 5-20 and 5-21).

For those special jobs, you can have your local wire EDM shop cut you special profiles and difficult-to-grind geometries.

Figure 5–20 This form tool was plunged into 4140 steel.

Figure 5-22 shows a couple of examples we have done over the years. Some would be a nightmare to hand grind accurately. Try finding commercially-made tooling for some of the crazy stuff designers come up with.

Figure 5–21 Another look at the same form tool.

Figure 5–22 Special tool bit profiles.

Figure 5–23 Backing tools in the Z axis.

Pull back on the tool post with a few pounds of force when backing your tools in the Z axis (Figure 5-23). This prevents leaving tool tracks in your turned surface. This works with boring bars on the ID also, but you have to push instead.

Save on holding stock by turning double ended parts, then cutting them off at the centerline (Figures 5-24 and 5-25).

Figure 5–24 Making double ended parts.

Figure 5–25 Cutting double ended parts at the centerline.

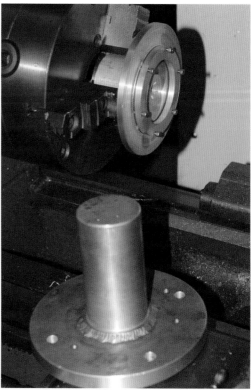

Figure 5–26 Aluminum face plate for a three-jaw chuck.

Make yourself a couple a quick aluminum face plates that fit in the three-jaw chuck (Figure 5-26). These can be re-surfaced dead flat dozens of times and are quick to set up. These two are welded together, but they can just as easily be bolted. If you bolt them, be sure to sink the heads well below the surface so you don't face the screw heads.

A nifty hand-tapping guide that fits in the tail-stock chuck works well for hand or slow speed, hand-tapping small thread sizes (Figure 5-27).

Figure 5–27 A hand-tapping guide.

Figure 5–28 A three-jaw backing plate.

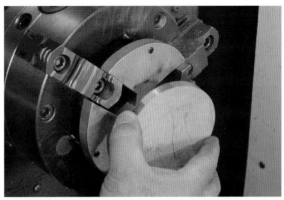

Figure 5–29 Positioning thin disc-shaped parts.

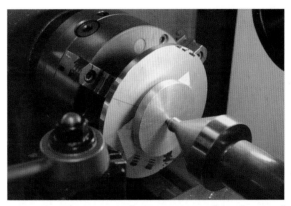

Figure 5–30 Pressure plating against a face plate.

Figure 5–31 Getting help from double stick tape!

Figure 5–32 Repetitive eccentric turning.

Three-jaw backing plates are great for backing up a part for heavy drilling (Figure 5-28). If you make several thicknesses, they can be used to position thin disc-shaped parts quickly (Figure 5-29). Add three-jacking screws for adjusting the plate in relation to the chuck face to hold thin disc-shaped parts right where you want them in the jaws against a nice parallel surface.

You can use light cuts for a disc-shaped part by pressure plating against a face-plate (Figure 5-30). You can even use the top of the chuck jaws if you have a little help. Three pieces of double stick tape make all the difference (Figure 5-31). Open the jaws to a radius just below what you will be turning to for maximum holding. Use a smaller disc with a center hole in it to push against with the tailstock center (Figure 5-30).

This method works great when you can't have a center drill or center mark in the work piece. Double stick tape always works better when the mating surfaces are cleaned with alcohol and the blank is squeezed into the tape.

For repetitive eccentric turning, clamp a small three-jaw chuck or collet block in your four-jaw chuck (Figure 5-32). Nobody likes to dial in every single part in a four jaw.

Figure 5–33 Boring tapers.

Figure 5–34 Running the lathe in reverse.

Turn the ID and OD taper with the same compound setting when trying to match tapers exactly (Figure 5-33). Run the lathe in reverse and cut on the backside for one of the tapers (Figure 5-34).

When turning tapers with a specific end diameter, make a couple of quick gages. Taper diameters are difficult to measure accurately in the machine. A little trig and a simple ring gage will make your life easier.

Use your flexible indicator holder on the top of the lathe tool post (Figure 5-35). There is just enough of a lip for the clamp to grab.

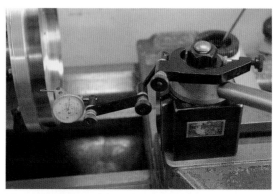

Figure 5–35 Flexible indicator holder.

A test indicator on a compact magnetic base is quick to set up, which makes you use it more often (Figure 5-36).

Hold large square parts outside the range of collets in a round sleeve (Figure 5-37). They are easy to make and are faster than a four-jaw chuck if you have a few square-to-round parts to do (Figure 5-38). A real organized person may make a few of these up for common square sizes.

Figure 5–36 A test indicator.

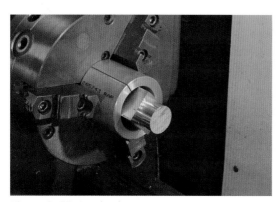

Figure 5–37 A split sleeve.

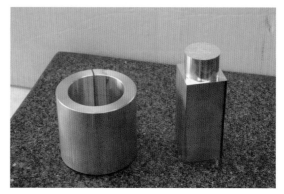

Figure 5–38 A split sleeve for squares.

Figure 5–39 Adding a short counter-bore.

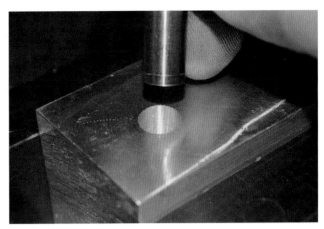

Figure 5–40 Turning the male part down.

Add a short counter-bore .001-.002 larger than the size of the part to be pressed in Figure 5-39. This makes assembly easier because the part is started dead straight. An alternate is to turn the male part down for a perfectly straight start (Figure 5-40).

Alternate lathe uses. You can press bearings and bushings in with the lathe tailstock (Figure 5-41). They go in straighter when pressed this way. Don't worry! The lathe and the mill can handle the thrust (Figure 5-42). What do you think happens when you're leaning on a big drill?

Figure 5-41 Pressing in bearings and bushings.

Figure 5–42 The lathe and mill can handle the thrust.

Align your parting tool against the flat face of the chuck (Figure 5-43). You can see tiny variations easily and it's super quick. The quick change tool-post can be aligned in the same manner.

Try power feeding your parting tool. Make sure it's sharp and square with the lathe axis. Keep it wet and start out with .002/Rev feed. The constant steady feed pressure works wonders. Most people have had trouble with parting at one time or another, so they tend to run too low a cutting speed and feed lightly.

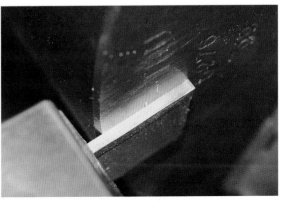

Figure 5–43 Aligning the parting tool against the chuck.

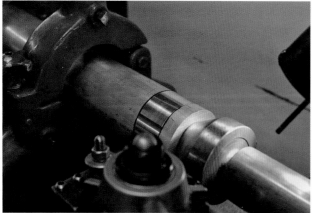

Figure 5–44 Turning true bands on tubes and shafts.

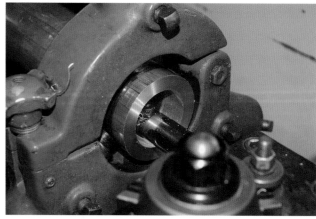

Figure 5–45 Getting bands to run true in a steady rest

For large diameter parting, you can part in several steps by moving the blade out in increments in the blade holder. Don't forget to adjust your center height when you move the blade in or out.

Turn true bands on tubes and shafts to get them to run true in a steady rest (Figures 5-44 and 5-45). If you don't turn a concentric band, the steady rest will just follow the out of round of the tube or shaft, producing a new surface to match. If you cannot cut a true band on the outside because of the final part geometry, you can clamp a sleeve on the outside and turn that true for the steady rest to run on.

When turning small part diameters, make sure your tool is dead on center and razor sharp.

Put a disc or a spider in the end of a tube that's too big for the steady rest (Figure 5-46). A bronze bearing and a shoulder bolt or a dowel pin in the drill chuck provides the pivot (Figure 5-47). This allows you to face the entire end of the tube square supported on center. You can part off big rings using this trick also. Just be sure you're parting off on the right hand side of the spider.

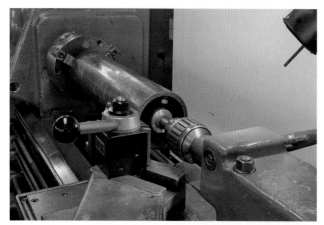

Figure 5–46 Putting a disc or a spider in the end of a tube.

Figure 5–47 Spider in the tailstock.

Figure 5–48 The quick vise grip method for springs.

Figure 5–49 A wire tension block.

When winding springs or rings in the lathe, you can use the quick vise grip method (Figure 5-48) or, if you want more consistent results, make a tension block that fits in the tool post (Figure 5-49). The factors that affect finish ring or spring diameters are mandrel diameter, wire condition, and initial tension. The vise grips are special homemade with the teeth removed and smooth copper jaws silver soldered in place (Figure 5-48). You don't need that much clamping force to hold the wire in tension as you wind it. Run the lathe at its slowest speed. Then be sure to engage the feed before you start the spindle when using the tension block. The wire centerline should be at the lower tangent point of the winding mandrel. Did I mention to make sure you run the lathe slowly? Turn the spindle on before you try this, to confirm the speed setting.

Figure 5–50 Squeeze the coil together.

Figure 5–51 Twist the legs gently into alignment.

Squeeze the coil together and cut all at once for the best results (Figure 5-50). Twist the legs gently into alignment and weld for professional results (Figure 5-51).

I keep several test indicators set up differently (Figure 5-52). You can jump from the height gage to the drill chuck to the spindle nose quickly without screwing around with all the little pesky clamp fittings.

Figure 5–52 A variety of test indicators.

Figure 5–53 A spin handle.

Figure 5–54 Using the chuck jaws on the lathe.

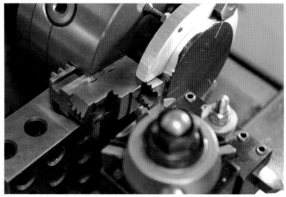

Figure 5–55 A well-placed block makes an accurate stop.

Make a spin handle for the compound rest if you hate turning short tapers as much as I do (Figure 5-53). You will get better finishes on your compound tapers and save your wrists on the heavy feeding.

You can use the chuck jaws on the lathe for a quick, accurate 120-degree layout (Figure 5-54). A block placed between the jaw and the way makes an accurate stop while you scratch a line with a pointed tool (Figure 5-55).

If you need multiple start threads, pick three-start if you have a choice. It's easy to index the shaft three times using the jaws of a three-jaw chuck as a reference surface.

Simple drive dogs can be quickly made from stock shaft collars (Figures 5-56 and 5-57). Turning between centers is still the most accurate way to establish concentric diameters end to end (Fig 5-58).

Figure 5–56 Simple drive dogs.

Figure 5–57 Using stock shaft collars.

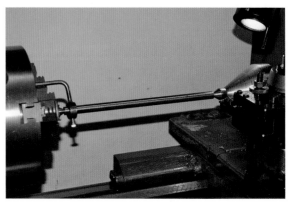

Figure 5–58 Turning between centers.

Figure 5–59 Turning a soft center.

Figure 5–60 Catching small parts.

Turn a soft center held in the three-jaw at 60 degrees included angle (Figure 5-59). It will be dead accurate on center if you don't move it after you cut it.

Set up a simple stop pin in a tool-block to save all your tool offsets on the DRO. Pull your stock out of the headstock against this pin to accurately set your starting position.

Put a small rod in the drill chuck to catch multiple small parts when parting off (Figure 5-60). It beats scratching around in the chip pan for your parts.

If you have a job where it's not an option to lose the part, do yourself a favor and clean the machine

Figure 5–61 Measuring the spindle centerline.

and chip pan before you start. You can also lay a sheet of brown paper in the pan so you can easily see what falls in.

Setting center height. Keep a square bar near each lathe long enough to span across the apron and accurately measure the spindle centerline distance off the bar (Figure 5-61).

This distance is permanently engraved on the bar. It makes it easy to set every kind of tool height no matter where the cutting edge is. On most lathes, the apron surface is the same height on both sides of the tool-post.

As an alternative, you can use the Sharpie centering method (Figure 5-62). Leave a tiny little spot on the face of a spinning part and set your tool to this target spot (Figure 5-63). Presto! Instant center height!

Figure 5–62 The Sharpie centering method.

Figure 5–63 Using a tiny spot as the target.

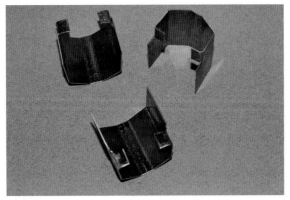

Figure 5–64 Copper sheet metal jaw covers.

Figure 5–65 Preclamping the jaws.

Make yourself three copper sheetmetal jaw covers for delicate work (Figure 5-64). Be sure to preclamp the new jaws on a test piece to set the final curvature (Figure 5-65). Safety wire or nylon ties keep them from dropping out every time you open the chuck. Or you can bend ears over like these. Material is .03 copper.

You can plunge turn small, long-aspect ratio diameters by taking the entire cut in one shot (Figure 5-66). Be sure the starting stock diameter is large enough to handle the extra cutting pressure without deflection.

You can also step turn this kind of part. Your chuck or collet must run very true for best results. The shaft is pulled out a little at a time as the diameter is reduced in steps along the z axis. This works best if the stock material is centerless ground for low runout.

Make yourself a retractable tailstock stock stop (Figure 5-67). When you have a bunch of parts to part off to a specific length, this works great.

Figure 5–66 Taking the entire cut in one shot.

Figure 5–67 Retractable tailstock stock shop.

When you need to cut a dovetail o-ring groove, set your compound at the half angle of the dovetail and cut along the angle with the compound (Figure 5-68). This allows you to use a larger, thicker tool than if you plunge in straight perpendicular moves for a job that on a good day can be difficult. The usual angle off the lathe centerline is 24 degrees.

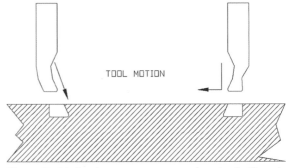

TOOL MOTION

Figure 5–68 Setting your compound at the half angle.

Figure 5–69 Clamping the rod.

Figure 5–70 Locking the bushing into the spindle bore.

You can use stock stop in headstock for repetitive chuck work. Clamp the rod with the screw in the middle of the bushing (Figure 5-69). The set screws in the counter-bore lock the bushing into the spindle bore (Figure 5-70). It uses a cut-down allen wrench to get at the screws. I don't really like this approach, but it does work and it is used so infrequently in our shop that nobody wants to spend the time to make a nifty one. Ideas, anyone?

Long stock guide. Whenever you have long stock sticking out of the headstock (Figure 5-71), you need to be extremely careful. It's best to cut the stock so it fits within the spindle.

But, if you must have it stick out, use a guide and run the spindle SLOWLY.

Much shop embarrassment, injury, and dental work have been caused by this specific operation. I have personally seen 2.5-diameter plastic whipping around at 3000 rpm and actually break off. Talk about run for cover. Check your spindle speed setting *before* you put the stock in. Did I mention this is a squirrelly operation?

Wind up a piece of thin, soft plastic sheet to keep the bars from getting scratched on the inside of the stock support tube (Figure 5-72). The plastic springs open when it's inserted in the tube, which helps retain it while the work piece turns inside. And for goodness sake, run the machine slowly!! Did I mention this is a very dicey operation and should only be used if you are in a bind?

Figure 5–71 A long stock guide.

Figure 5–72 A plastic sheet protects the bar.

Figure 5–73 Soft jaws.

Figure 5–74 Boring sprockets.

Here are some really easy-to-make soft jaws that can be rotated and have four usable edges (Figure 5-73). Use these for light-to-medium work, like boring sprockets accurately on the pitch circle (Figure 5-74).

Make yourself a couple of soft jaw spiders from large hex nuts (Figures 5-75 and 5-76). They will cover a million diameters and save hunting down the right disc in the scrap bin. You can easily set them with calipers (Figure 5-77).

Figure 5–75 Soft jaw spiders.

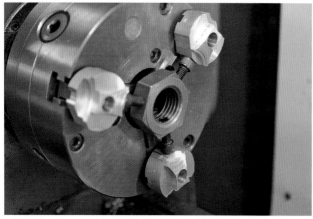

Figure 5–76 Using large hex nuts.

Figure 5–77 Setting soft jaw spiders with calipers.

5.3 Step Turning

Step turn large radii in the manual lathe using x and z coordinates. You can cut large radii in the lathe using a method I call step turning. If you don't have a CNC lathe, this might be the only option for large curves. This is one of those things that is much easier to do with the assistance of a computer for the graphic layout and all the math. You can do the math manually, but it takes a little longer.

The first step is to lay out the desired curve accurately at full

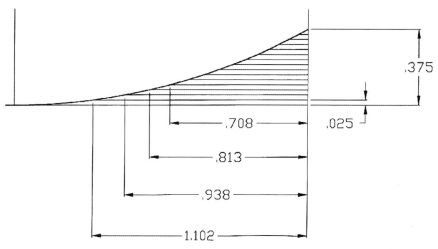

Figure 5–78 Cutting a large radius.

scale or, if you're doing the calculations manually, maybe ten times scale for more accurate final results.

In this example, we are trying to cut the large radius in Figure 5-78. We can do this by taking a series of cuts of a certain depth that all have their z endpoints on the desired curve. In this case, we are taking a radial depth of cut of .025 per pass. Each successive pass has a different z end point that ends exactly on the curve. Depending on the accuracy needed, you can take as many passes as you like. Don't make the radius diameter mistake. In the above example, we are only showing half the curve, but removing .050 on the diameter per pass.

Figures 5-79 and 5-80 show a lower anvil for a large English wheel machine. You can see both side have been step turned and are ready for the finishing. If you're doing the layout manually without the benefit of a computer, measure your points on the curve from a baseline with calipers. It helps to draw your curve 10x scale for better accuracy.

Figure 5–79 Lower anvil of a large English wheel machine.

Figure 5–80 The sides are ready for finishing.

Figure 5–81 A partially blended curve.

Figure 5–82 Removing excess material is the next step.

In Figures 5-81 and 5-82, you can see where the curve is partially blended in. We already know the corners of the steps lie directly on the actual curve, so all we need to do is remove the excess material between each pass. If you take a large number of passes, the material left to remove can be as small as you like. Also, the trailing edge of the tool can be shaped to help remove the excess between the step lines.

You can use a rotary type radius tool in the lathe instead of grinding one from scratch (Figure 5-83).

Figure 5–83 Using a rotary type radius tool.

5.4 Threading in the Manual Lathe

Screws and screw threads literally hold the world together. There are as many types and forms of threads as there are products that use threaded fasteners and connections. There's also confusion and misuse of threads in general for the non gear-head genre.

From the machinist point of view, cutting threads is kind of a satisfying experience. When you're done, hopefully you have two parts that mate together with a precision and smoothness you just don't get with run-of-the-mill hardware grade fasteners. I have always enjoyed cutting threads in the manual lathe and have learned a few tricks over the years that are of interest.

Align your tool against a freshly-faced end or against the side of the chuck (Figure 5-84). The little arrow-shaped alignment tools you see are a pain and are only good for gaging hand ground tool bits.

Figure 5–84 Aligning your tool.

Figure 5–85 Starting your threads.

Figure 5–86 Chamfer with the threading tool.

Figure 5–87 Traversing a small relief at the end.

Figure 5–88 Switching to a radius tool.

If you do a lot of threading in the manual lathe, invest in an inserted tool-holder. The inserts are ground with near perfect geometry and are easily changed. One insert cuts dozens of thread pitches.

When I learned how to thread in the lathe, I learned using the compound in-feed method. Contrary to some popular belief, you do not have to have the compound set at half the thread angle. By using what's called modified flank in-feed, and changing this angle, you can help alleviate threading problems with difficult materials.

Another advantage to threading with the compound is you don't have to keep track of your dial position. The cross-feed dial is always zeroed after each pass, so you have less to remember. For example, was that last pass at .030 or .050? The main disadvantage is your Z position changes as you feed in. This is usually not a problem on OD threads, but it can be a problem on internal threads that end against a shoulder.

How do you start your threads? Figures 5-85 and 5-86 show a couple of examples. I like the chamfer with the threading tool (Figure 5-86). It saves a tool change. Be sure to chamfer a little deeper than the minor diameter of the thread.

How do you end your threads when the designer has not specified a specific relief? Here are a couple of ideas to try out. When I want to do something with the groove that gets cut at the end of the thread, I usually just use the threading tool and traverse a small relief at the end (Figure 5-87). It saves a tool change and looks okay. If I want a little nicer look, I sometimes switch to a radius tool (Figure 5-88). Just be sure you are a little smaller with the relief than the thread minor diameter so the mating part will thread all the way to the shoulder.

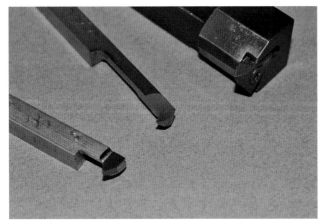

Figure 5–89 Left-handed tools.

Figure 5–90 Internal threading from the inside out.

Use a large DOC on your first pass during threading. The point is small; in the first couple of passes, the area of the tool tip engagement is also small. Taper off your depth of cut as you get deeper.

On your last pass, feed straight in with the cross feed a light .001 spring cut. This cuts on both flanks of the tool and literally cleans the thread of any chatter or tool marks.

I can never remember which line on the threading dial to use with which thread pitch. If you're lucky, it will be marked. When in doubt, just use the same number or line each time. Always use the same number when cutting multiple start threads.

Do your internal threading from the inside out with left-handed tools (Figures 5-89 and 5-90). You will get less chatter and you can see what's going on down the bore. You will need left-hand threading tools and run the lathe in reverse. The way it was explained to me was, "It's easy to pull a rope; it's really hard to push one."

When you have a choice, fine threads are easier to cut and need fewer passes. The shallower depth on difficult materials might save your bacon.

For quick and easy day-to-day threading gages, I keep a complete set of nuts in my toolbox for fitting threads (Figures 5-91 and 5-92). One ring hold coarse threads and the other ring holds fine threads. When you thread, be sure to run the nut the full length of the threads. When left to their own devices, machinists tend to cut threads tighter than necessary.

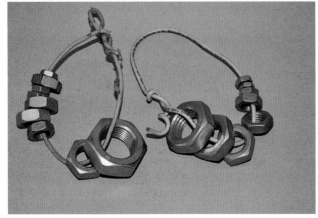

Figure 5–91 Keep a complete set of nuts.

Figure 5–92 Run the nut the full length.

Mating materials in threaded connections are important. If you must use the same material for male and female threads, do yourself a favor and put a few molecules of thread lubricant or anti-seize on them before you crank them together.

If you do happen to get your male and female threads wedged together in an intimate embrace, a simple trick to get them separated is to quickly warm the female portion 100 degrees or so, using a propane torch. A quick shot of penetrating lube before you twist and you might just save your work.

When measuring threads, a dedicated thread micrometer is very handy and quick at the machine. But for the highest accuracy, use the three-wire thread measuring method. The reason the three-wire method is more accurate is because the wires present a true parallel surface for measuring. If it's good enough for the gage makers, it's good enough for me.

A piece of modeling clay or window glazing putty can help you hold those pesky thread measuring wires. Better yet, buy a set of the plastic holders that fit the micrometer spindle.

Thread files actually work (Figure 5-93). They are great for straightening the annoying half thread fade at the beginning and end of an external thread.

Figure 5–93 A thread file.

5.5 Multiple Start Threads

Here is an example of cutting multiple start threads, sometimes called multiple groove threads, in the manual lathe. These threads are used for getting a high lead per revolution with a shallow thread depth.

Suppose you have a .25-inch per revolution lead, but it is to be cut on a small diameter cylinder or thin-walled tube. The normal double depth for a .25-lead, 60-degree thread is .324. If you wanted to cut this on a shaft of .375 diameter, you would pretty much be out of luck. Enter the multiple start thread.

As the name implies, there is more than the normal single start. These threads can be identified by looking at the end of the thread and counting the number of entry starts you see. There is no practical limit to the number of starts you could do. The limitations typically are with the machinery used to produce them. Most engine lathes will not thread coarser than 2 TPI. In the CNC lathe, this is not the case, but that's another chapter.

Let's work through the example I have done. The process is the same as cutting normal 60-degree threads we find all over the place with a few exceptions.

For multiple start threads, we must index or adjust our starting position for each separate start. This can be done in several ways. The first way is to index the part radially precisely the number of starts you wish to have. So if you have a three-start thread, you would index each start 120 degrees. Or if you wanted a four start, you would index your part ninety degrees. It's important to note that the axial or Z position cannot change at all when you use the part index method. This limits you to threading the part between centers to maintain the same Z position.

Figure 5–94 Prepare the threading blank.

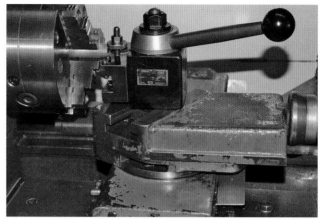

Figure 5–95 Prepare the levers, dials, and compound.

The second method is easier, but only if your lathe has a compound rest attachment which will swing ninety degrees to be aligned with the bed of the lathe.

The first step is to prepare your threading blank. In the example, the diameter is arbitrary (Figure 5-94). We will cut a four-start thread with a lead of .25 per revolution. To determine the actual thread dimensions to cut the thread, we will divide the .25 lead by the number of starts. So we have (.25/4) = .0625. This corresponds to a thread with a lead of one-sixteenth per revolution or 16 TPI. This is the thread depth we will cut for each of the four starts.

Make sure your threading levers are set for the desired lead which, in this example, is .25 per revolution or 4 TPI. Remember we are cutting a 4-TPI thread, but with four individual 16-TPI depth grooves.

I always start by zeroing all my dials cross feed and compound. Make sure the compound is set at 90 degrees and the dial is zeroed (Figure 5-95).

Out of old habit, I always take a .001 scratch pass to confirm my gearbox settings (Figure 5-96 and 5-97). You would be surprised how many times this gets fouled up. For high lead threads, you will want to run the spindle pretty slowly, especially if you are heading toward a shoulder like this example.

Figure 5–96 Make the first scratch.

Figure 5–97 Confirm the settings.

Figure 5–98 The thread dial.

Figure 5–99 The first groove at final depth.

Figure 5–100 The second groove at scratch pass.

Figure 5–101 The second groove at final depth.

I always use the same number on the thread dial for multiple start threads (Figure 5-98). It is most likely okay to use other lines or numbers, but who wants to screw up their work? Typically threading is one of the last operations, so you have invested some time to get this far. Why risk a failure? In Figure 5-99, I have cut the first groove to final depth.

Now for the second groove. The compound rest must be moved a distance along the z axis that is equal to the thread lead divided by the number of starts. In our example, (.25/4) = .0625. So we advance the compound .0625. The direction that the compound is moved does not matter as long as you don't change once you start in a particular direction.

I made a confirming scratch pass for the second groove to be sure I moved the compound the correct amount (Figure 5-100). In Figure 5-101, the second groove is cut to full depth. After completing each groove, the compound is advanced the distance of (lead divided by number of starts) along the z axis.

Figure 5–102 The third groove.

Figure 5–103 The fourth groove.

After the third groove, it's starting to look like something (Figure 5-102). And after the fourth groove, it looks like a funny 16 TPI (Figure 5-103). In fact, if you put a standard thread gage in the grooves, it should fit the 16 leaf. The only thing that looks different is the lead angle, which looks much steeper than a normal 16 TPI.

The process is fundamentally the same for an internal thread. If you find yourself having to do an internal multiple start thread, be sure to try the inside out ID threading that was discussed earlier. When you make the mating female thread for this example, it will have a lead of .25 per revolution, but a thread depth of a 16 TPI. Pretty cool!

Figures 5-104 and 5-105 show a pretty interesting device I saw recently on an old lathe. It's a hinged, flip-down drill chuck attached to the apron. It has a clever latch to lock it into place on the machine centerline. This allows you to use the power feed for big holes and the hand wheel for rapid pecking on deep holes. It also allows you to have more than one centerline tool set up at once like a drill and a tap.

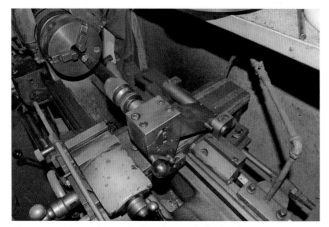

Figure 5–104 A hinged, flip-down drill chuck.

Figure 5–105 The chuck is attached to the apron.

Figure 5–106 An indicator coupled to the tailstock quill.

Figure 5–107 A tiny Manson pocket backpacking lathe. Just kidding; it's not for backpacking!

Here is a way to quickly couple your indicator to the tailstock quill (Figure 5-106). Sometimes it's important to get a precision depth with the tailstock. This method saves making a special bracket that most likely only fits one lathe. You can count turns if you like, but I find this method less prone to daydreaming.

6.1   Bridgeport Mills
6.2   Suggested Improvements
6.3   Spherical Surface Generation

Manual Milling Machine

If the manual lathe is the king of machines, then the manual mill must be the queen. A shop is just not a shop without at least one manual mill. Years ago, before the current vertical mill configuration became readily available, the shaper was the shop heavyweight for prismatic type work. Ever since the introduction and evolution of the modern vertical milling machine, most machinists today could not even set a shaper up, let alone get some work done.

6.1 Bridgeport Mills

When everybody thinks of the vertical milling machine, the first thought is a "Bridgeport." This is the machine by which all others are judged. Now I have run a few Bridgeports in my time and I would have to say they are a nice machine. They have the look and feel of a well-made machine. All the levers turn and lock smoothly with just the right click and feel. The height, width, and depth lend themselves to machinists of average height and reach. They are also far from perfect.

What happened to the Bridgeport mill often happens to an average design that meets high demand and sales. The design stays static and all the design flaws are faithfully reproduced in the army of clones marching out of the factory. Worse yet, the competitors who have a chance to correct the problems instead copy the flaws in their own brand of knockoffs.

I can also say that if I had been the president of Bridgeport, I would probably have done the same thing. Why mess with success? Bridgeport had the manufacturing capacity and most evolved design at the time when there was a high demand for machines.

The only reason I even bring it up is so designers and machinery builders can learn and hopefully advance the state of tool design to the next level. The design should never have become static in the first place. Anybody who has spent time on a Bridgeport or clone of one will be able to relate to the basic design flaws.

The *Y*-axis dovetail ways are much too narrow in relation to the length of the table. The entire *X*-axis can be rocked back and forth when the gibs are not set snugly. Set the gibs too tight for minimal play and your arm is dead at the end of the day.

The *Y*-axis ways should be extended to double their current width. This has the added benefit of covering and protecting the exposed ways behind the table where all the chips land and damage the

ways. If you have ever looked at a clapped out vertical mill, this is one area where they show their age.

The head tipping feature is grossly offset from the center of gravity of the head. Heavy cuts can easily knock the head out of tram. The pivots should be on the centerlines of the spindle at least. At the very least, eliminate to the front-to-back tilt feature and go with a single right-left tilt.

Figure 6–1 A Bridgeport vertical milling machine.

6.2 Suggested Improvements

The entire drawbar assembly could be improved drastically. Think about tool retention in CNC equipment. Thousands of lost man-hours could be mined with a few simple improvements in this area.

Quill locking feature. I have seen quite a few novel methods to keep this limp locking device from dragging. They are all weak band-aide fixes for a bad design.

Axis locking screws. These tend to push the axis off position when activated. How about some blade type locks as per jig borer design?

Acme lead screws. Excellent ball screws have long been available. When will manual machine manufacturers take advantage of these low-friction enhancements?

So, until some clever person decides to really take an objective look at the vertical turret milling machine and make some changes, we're all stuck with what we have. Most machinists are pretty clever themselves. There is more than one way to make one of these machines make parts and money.

Okay, on to the manual mill section!

Tramming and aligning. Get yourself an indicator with a vertical dial (Figure 6-2). This beats craning your neck like a bird hunting a worm all around the mill when zeroing a part in. For tramming the head in, I made several long indicator holder bars to sweep a larger arc (Figures 6-3 and 6-4).

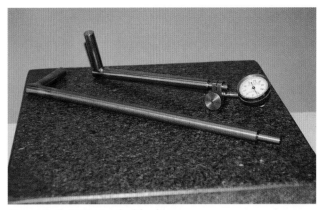

Figure 6–3 Indicator holder bars.

Figure 6–4 The longer bar sweeps a larger arc.

If you are having trouble indicating a bore that you think should be round, be sure to check your head tram condition. Typically you would see a longer direction or equal but mismatched sweep numbers. If the head is out, you're out on your head.

Button type indicators slide over the tee slots easier than a test indicator when tramming (Figure 6-5).

Figure 6–2 An indicator with a vertical dial.

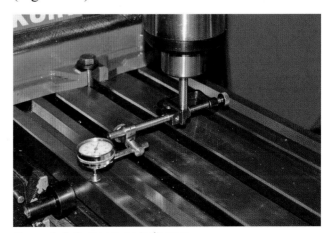

Figure 6–5 Button type indicators.

Figure 6–6 Using a combination square.

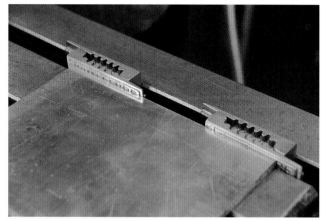

Figure 6–8 Two wedges set into the tee slots.

Use your high-quality combination square on the mill (Figure 6-6). You will be surprised how close you can get. This saves decades of indicating time. I checked one of my combination squares against a master square comparator recently and was happy to find that it was within .002 of square at ten inches off the surface plate.

Figure 6-7 shows special toothed wedges I use as backstops in the mill. I used to use 5/8 dowel pins like everybody else until I found these. These wedge tight in any width tee slot. You can vary the height easily and even use them for special clamping jobs.

Figure 6-8 shows two wedges set into the tee slots as a backstop for a plate. These unique wedges are called Quoins. They were used in the printing industry to keep type securely into the type case. As usual, I have bent them into service for a different task.

Drilling chucks. Cut the stupid long shanks off your drill chucks (Figure 6-9). The R-8 collet is only an inch long inside. How much shank do you need up there? Think about all the time you spend cranking the knee up and down to get the chuck in and out with that long shank.

For that matter lose the R-8 shank on your drill chuck. If you have a 5/8-diameter straight shank on your drill chuck, you will save hundreds of collet changes a year if you buy a few common sized end mills with the same shank size as your drill chuck. That's in addition to all the time wasted cranking the knee.

Buy a couple of drill chucks. If you have different diameter shanks on them, you can save time on tool changes when you have the same shank diameter as your cutting tools.

Figure 6–7 Special toothed wedges used as backstops.

Figure 6–9 Cut long shanks of the drill chucks.

Knee Crank Trick

This is an old shop trick to be used on apprentices and newbies. Everybody hates manually cranking the knee up after having near the bottom of the travel. The trick goes like this. As the hapless victim walks nearby, pretend to be listening as you slowly crank the knee up. Give a little harrumph of concern just as they get close. If you're lucky, they will ask what's wrong. If not, call them over. Ask them to crank the knee up and see if they hear the noise. You will need to be non-specific about the actual noise. As they crank it up, listen and comment appropriately. "There it is! Did you hear that? Crank it again, a little faster, that's it. Did you hear it that time?" When you reach the desired height, shake you head in mock concern and say something like, "We need to keep an eye on this machine" or something along those lines.

Don't put end mills in quick-change drill chucks—though tempting, it's a pure rookie move. If the end mill chatters for a billionth of a second, the chuck loosens; all hell breaks loose. I saw somebody do this one time....

Wear out a drawbar once in a while. Drawbars are cheap compared to work that's spoiled because the collet wasn't tightened enough.

Drawbars. Remove the drawbar once in a while and put a drop of oil or light assembly lube on the threads. You should be able to spin this with your blistered pinkie finger. If it doesn't spin freely, get a new one.

Use the spindle motor to rapid traverse the collet out once you have it broken loose with the wrench. Only hold the drawbar with your fingers lightly and catch the collet as it falls out. Never use the wrench.

Better yet, invest in a power drawbar (Figure 6-10). If you haven't tried one of these you should. This device has a very short payback period. Unless you use your right angle head on a daily basis, the argument about slow changeover doesn't hold any water. Time a few tool changes and do some math to see what I mean. Most people would easily invest in a DRO for a manual mill for the convenience.

Once you have tried a power drawbar, you will wonder how you got along without it. No more smashed fingers or wrenches ratting around over you head.

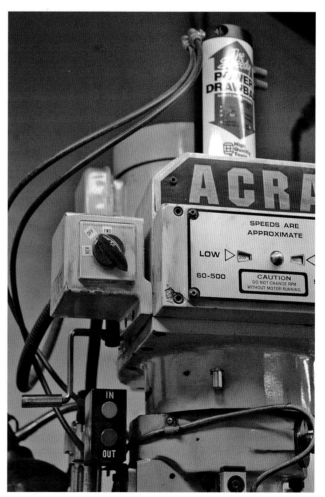

Figure 6–10 A power drawbar.

Figure 6–11 Hold your hand on the part.

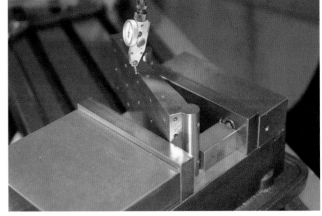

Figure 6–12 Setting the head or vise angles.

Hold your hand on the part when using tricky or dicey setups (Figure 6-11). Your hand will detect a part shifting before your eye will see it, giving instant feedback to the other hand that's cranking the feed handle.

Sine bars. Get used to using sine bars. These are simple-to-use, deadly accurate, angle-setting tools. Your sine bar should span across the ways of your standard milling vise. Don't think of them as too precise to use for everyday work. Smaller sine bars are handier for manual mill work. A three-to-five inch center distance is perfect.

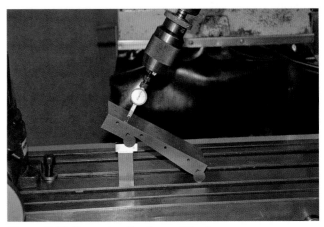

Figure 6-13 Sweeping the face of the bar.

Use your sine bar to set head or vise angles (Figure 6-12). You can also sweep the face of the bar just like you would when you tram the head to set a precision angle (Figure 6-13).

Do a sanity check with your protractor to confirm angle settings (Figure 6-14). I saw somebody make a little math boo boo once….

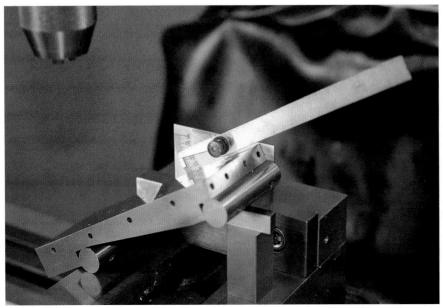

Figure 6–14 Confirming angle settings with a protractor.

You can use a gage pin (Figure 6-15) or an adjustable parallel (Figure 6-16) to set your sine bar quickly. This can save stack-up math errors with gage blocks. As a good measure, always caliper a stack of blocks to confirm your math (Figure 6-17).

Sometimes it's faster and easier to make a quick drill fixture to drill and tap holes in the edges of large plates(Figure 6-18) . The setup time and handling can be murderous for just a few holes.

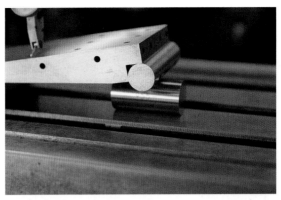

Figure 6–15 Using a gage pin.

Figure 6–16 Using an adjustable parallel.

Figure 6–17 Confirm math with calipers.

Figure 6–18 Drilling and tapping holes.

Figure 6–19 Make the block an accurate size.

Make the block an accurate size so you can locate it precisely on the plate (Figure 6-19).

Figure 6-20 shows a technique that beats setting up the right angle head or hanging the plate off the side of the mill. If you use a hardened drill bushing, the hole comes out straight and in the correct position. You can even make a second fixture for tapping the holes if the additional time is warranted.

Figure 6–20 Another technique for drilling a hole.

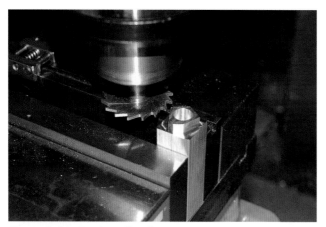

Figure 6–21 Cutting off mill parts.

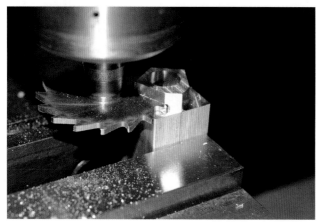

Figure 6–22 This technique works well for very small parts.

Cutting mill parts. Mill parts complete. Then cut them off with a slitting saw or key-seat cutter (Figures 6-21 and 6-22). This works particularly well for very small parts. Be sure to keep your eye on the part when it comes off or you might spend more time looking for it than it took to make it.

You can use your right angle head like a precision cold saw to neatly cut parts accurately to length (Figure 6-23). This is a good use of a retractable stop to eliminate parts jamming between the stop and the blade. Use blade with enough thickness to cut straight.

Figure 6–23 Using your right angle head.

You should be able to cut parts within a couple of thousandths (Figure 6-24). Make sure the bottom of the right angle head clears the part and the vise (Figure 6-25). Also be sure to retract your stop before the part comes off or it may jam.

Figure 6–24 Cutting parts precisely.

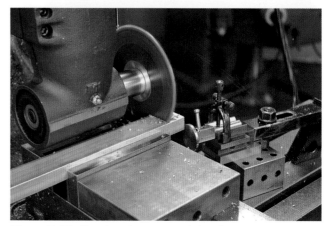

Figure 6–25 Clearing the part and the vise.

Figure 6–26 A knee rapid feed.

Figure 6-26 shows a poor man's knee rapid feed. This will save you quite a bit of time over the course of a year. It has wrench flats on the shank so you can still make fine adjustments with a wrench. Nobody can say I'm **not** a cheapskate.

Right angle and boring head tricks. You can quickly align the right angle head if your vise is straight with the world. With the clamp screws for the right angle head lightly clamping the body, drop the head into the jaws of the vise and snug the vise slightly (Figure 6-27). For fussy work, you will still want to indicate the head for perfect alignment. By the way, don't trust the flats on the side for fussy work. Indicate a test bar held in a collet.

Cut odd radii with a boring head (Figure 6-28). This is a tube-bending die for an odd centerline radius. The part is rotated by the rotary table with the boring tool cutting edge on the centerline (Figure 6-29).

In a pinch, you can cut common angle chamfers with a standard single or multi-flute countersink. Stay off the small diameter tip for best cutting results. The tip edge has insufficient chip clearance to cut a chamfer.

If you place a piece of brown wrapping paper under plates clamped to the mill table, you will never have a plate shift on you again. This tiny bit of paper acts like a brake lining to keep your part from slipping. It has the added benefit of protecting the finish on your parts from scratches from the table surface.

Figure 6–28 Cut odd radii with a boring head.

Figure 6–27 Dropping the head into the jaws of the vise.

Figure 6–29 The part is rotated by the rotary table.

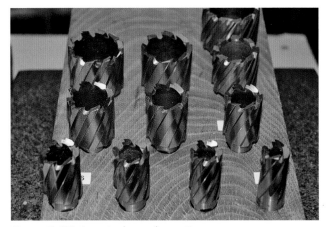

Figure 6–30 Assorted annular cutters.

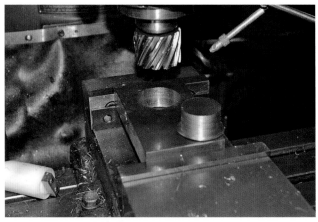

Figure 6–31 Slugs that can be re-used.

Annular cutters. Try using annular cutters (Figure 6-30) for hole making in the mill. They produce accurate holes and need a fraction of the feed pressure to make them cut. You also end up with a neat little slug that you can use for something else (Figure 6-31). Do not stack cut with them, period. Annular cutting tools are much more efficient than drill bits (Figure 6-32). The cutting speed of a normal drill bit approaches zero at the tip. Essentially the center portion of a drill bit is broaching it way through the material. Annular cutters maintain a more uniform cutting speed and convert more energy into hole and less energy into chips.

These cutters could care less if they cut a full hole or some fractional part of one (Figure 6-33).

This example shows a two-inch hole in one-inch steel. Hole time was less than one minute, hole accuracy +/− .002 The cutters can stay in the cut because the helical flutes on the outside extract the chips, unlike a holesaw.

Annular just means little ring. Annular cutters remove a narrow ring of material at the periphery, which is the secret to their high efficiency. They use less horsepower and less thrust to produce a given hole. These highly efficient hole-making tools are available to cut everything from sheet metal to thick cross section material several inches thick. An added bonus is the superior hole surface finish they produce. If used properly they will last for years and can be re-sharpened several times.

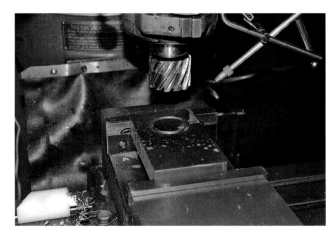

Figure 6–32 Annular cutting tools are efficient.

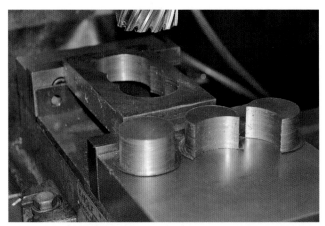

Figure 6–33 Annular cutting tools are versatile.

Figure 6–34 Cleaning the teeth of the hole saw.

Figure 6–35 Pre-drilling holes in the saw groove.

Hole saws. When using hole saws in the mill, follow these suggestions for excellent results. Only stay in the cut for two or three seconds at a time. Back the saw out after each three-second peck. After two pecks, clean the teeth of the hole saw with a small wire brush or air hose while it's running (Figure 6-34). This clears the chips out of the teeth and keeps it cutting. Another thing you can do is pre-drill one-to-four holes in the saw groove to break the chips and strip them out of the teeth of the hole saw (Figure 6-35).

Mill your blocks square with the end cutting surface of your tool. Less tool flex makes for better parts. Learn the 7-step method of squaring blocks without a tool change (see Appendix A).

Figure 6–36a Checking flatness of parts.

Figure 6–36b Using a three-point leveling system.

Check flatness of parts with a three-point leveling system (Figure 6-36). Level the underside datum to the same indicator reading. Then sweep the top surface, flip, and repeat. I bet you always wondered what those pointed screw tips were for. Another easy way to support parts for flatness checking is to use three precision balls as your part rest (Figure 6-37). Verify all three balls are the same size. You can use putty or modeling clay to keep the balls in position for repetitive part checking. The three ball method is faster because you don't have to adjust the part height when leveling.

Figure 6–37 Three precision balls with putty.

Figure 6–38 Using a V-block.

Figure 6–39 An extra tall V-block.

V-blocks. Use a V-block for holding round stuff in the mill vise (Figure 6-38). It gives three-point contact and automatically squares the stock to the jaw accurately.

Extra tall V-blocks can be set up with a couple of angle plates (Figures 6-39, 6-40, and 6-41). One advantage of doing this is you can set up the angle between the two plates to anything you want. Be sure to indicate the faces vertically to insure they are straight with the world. You can also leave a space between the plates for clamping purposes.

Figure 6–40 Adjusting the angle of a V-block.

Figure 6–41 Another V-block.

Or you can make a soft jaw that has a V-groove already in it for three-point holding a variety of round items (Figures 6-42 and 6-43).

Figure 6–42 Three points hold round items.

Figure 6–43 A soft jaw with a V-groove.

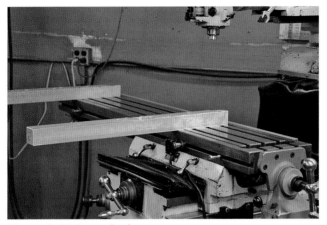

Figure 6–44 Spreader bars.

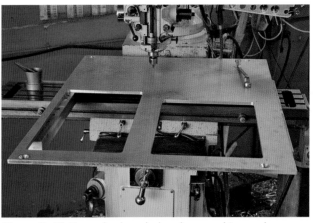

Figure 6–45 Spreader bars help when working on large plates.

More journeymen milling tricks. For working on large plates, make yourself a couple of spreader bars (Figures 6-44 and 6-45). The spreader bars have counter-bored bolt holes on the same centers as the tee slots. Large plates can be C-clamped to the overhang directly to the bars. Make the spreader bars sacrificial out of something soft.

Figure 6–46 Using a honing stone to round corners.

Figure 6–47 Improving the finish of drilled holes.

If you round the cutting corners of a standard drill bit with a honing stone (Figure 6-46), it will leave a better finish in the drilled hole (Figure 6-47). This is a sneaky way to get around an odd reamer size you don't have. You can also hand grind a twist drill a tiny bit off center to get it to cut a whisker larger.

With an oversize table on your vertical mill, you can have more than one setup on the machine (Figure 6-48). All our machines are oversize tables; the dovetails are wiped clean all the way to the ends of the travels. We have really made use of that extra work envelope.

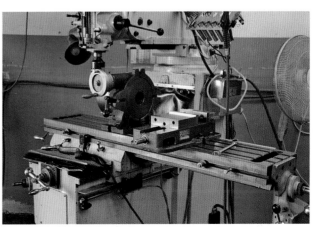

Figure 6–48 More than one setup on a machine.

Figure 6–49 A slip of paper helps with measurement.

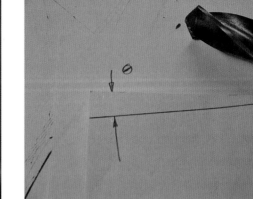

Figure 6–50 Forming the clearance angle.

A simple way to measure the back clearance or relief angle of a hand ground drill bit is to use a little slip of paper (Figure 6-49). Wind the paper around the cylindrical part or the drill body smoothly and mark the point where the corner meets the edge of the paper. The angle formed by these points is the clearance angle if you ever wanted to measure it (Figure 6-50).

Casting compounds. You can use castable urethane to support thin-shelled parts for milling or turning (Figures 6-51 and 6-52). This Delrin part would have been nearly impossible to machine with its .03 wall thickness without internal support. Use a little automotive wax or mold release so the part pops out.

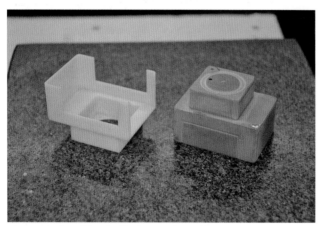

Figure 6–51 Thin-shelled parts for milling or turning.

Figure 6–52 Some parts need internal support.

Figure 6-53 shows another example of using a casting compound to hold or support a part that would be difficult at best to fixture quickly. The tube was held in a chuck in the mill and the casting compound was then poured in an old parts box and around the part to encase it.

Figure 6–53 This casting compound holds a part.

Figure 6–54 Splitting the block down the parting line.

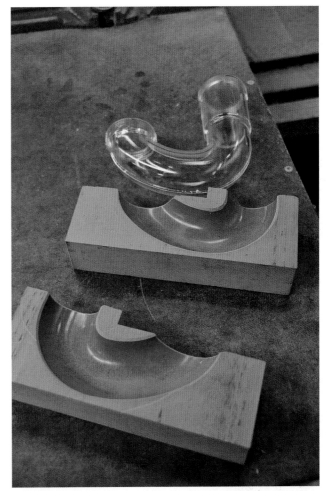

The outside of the casting compound was squared up. Then the block was split with a band saw down the parting line we wanted (Figure 6-54).

Now we have a perfect negative shape we can use to register additional parts for machining (Figures 6-55 and 6-56).

This urethane material called Pro-Cast sets up in about 20 minutes, making it a real handy trick for holding impossible parts for precision machining (Figure 6-57).

Figure 6–55 Creating a perfect negative shape.

Figure 6–56 The negative shape can register additional parts.

Figure 6–57 Pro-Cast holds impossible parts.

Figure 6–58 Cerrobend is a low-temperature melting metal alloy.

Figure 6–59 Wrapping with aluminum foil tape.

The next two figures provide another example of using a castable material to support a part for machining. Figure 6-58 shows a low-temperature melting metal alloy called *Cerrobend*. It melts at approx 150°F. The part is an aluminum heat sink where the thin fins were filled with the metal and then machined together. The wrapping is aluminum foil tape, which forms the dam to contain the liquid metal (Figure 6-59). Extra-melted material can be reclaimed and cast into pucks in a muffin tin for the next use. This material works best if not overheated from its liquid point very much. Oil the part before pouring to make the material completely melt out on removal.

Put a short wrench on the vise to keep apprentices from over-tightening mill setups (Figure 6-60). Use the clock method to describe the handle position. "Okay, only tighten this to three o'clock, got it? Not two-thirty and not three-thirty." This only works if they don't have a digital watch….We add the nylon tie to keep the handle from falling off the clock.

Figure 6–60 Using a short wrench to prevent over-tightening.

Figure 6–61 A handy mill fixture.

Figure 6–62 The part rests on the pins at the bottom.

Drilling and tapping. Figure 6-61 shows a handy mill fixture for drilling and tapping the ends of long bars. It bolts to the mill table and the mill head is rotated over to one side. On one side, the fixture has a little vertical fence to make sure the part is square. The pins at the bottom are what the part rests on to take the machining thrust (Figure 6-62). Around the back side there is a stiffener under the mill table. A small jacking screw against the dovetail bottom stiffens and squares the assembly (Figure 6-63). You can use your drill chuck as a hand tapping guide if you leave the collet loose (Figure 6-64). Don't forget to lift it up before you move to the next hole. I saw somebody do that once….

Figure 6–63 A small jacking screw.

Figure 6–64 Using the drill chuck as a hand tapping guide.

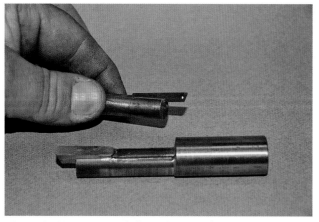

Figure 6–65 Homemade broaching tools.

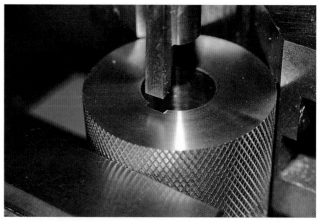

Figure 6–66 Cutting a blind keyway.

Broaching tools. Figure 6-65 shows homemade broaching tools for cutting blind keyways or shaving internal corners square in the milling machine. They cut like a vertical shaper might. The shanks are eccentric so the tool can enter the bore size and keep the cutting edge closer to centerline. Don't tell anybody you can do this or they will want it done all the time. It's kind of like being a dentist and scraping some tartar out of the garlic-eating champion's mouth. This is for special cases only. If you have to cut a blind keyway like this, put a relief at the bottom for the tool to clear the material and break the chip (Figure 6-66).

Chuck and vise jaws. Get a cheap air ratchet for changing chuck and vise jaws (Figure 6-67). I replace the mounting screws with shorter ones for less time spent spinning. Be sure to take any cutting tools out of the spindle. The air ratchet can buck your hand into the tool if you're not ready for the torque reaction.

Counter-bores. Figures 6-68 and 6-69 show an easy to make counter-bore made from a normal twist drill. It's handy for those odd sizes that come up when designers don't have a regular counter-bore chart. A quick spin on the surface grinder or Deckel tool grinder and off you go.

Figure 6–67 An air ratchet helps change chuck and vise jaws.

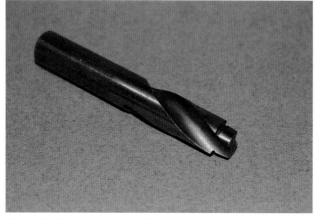

Figure 6–68 An easy-to-make counter-bore.

Figure 6–69 The counter-bore is made from a twist drill.

Figure 6–70 Making up small differences in part width.

Figure 6–71 Working with multiple stacked parts.

Securing parts. Try using some soft aluminum filler rod swiped from the welding department to help secure multiple parts. The soft round wire squeezes down and makes up the small differences in part width (Figure 6-70). This trick works best on hard parts like steel or stainless that you have a fair amount of clamping pressure. Put the soft wire in vertically for multiple stacked parts (Figure 6-71). The round wire deforms more easily than flat material because the pressure is on a point instead of a line.

Figures 6-72 and 6-73 show another great trick for holding multiple parts securely. This is a standard dovetail o-ring groove cut into the soft jaw face (Figure 6-72).

A cutoff piece or o-ring cord stock provides the right amount of squeeze and friction to hold multiple parts securely (Figure 6-73). You can also use different o-ring materials for different types of part holding.

You can insert a sliver of shim stock in one side of a collet to get an end mill to runout and cut a whisker bigger. This works when you need an end mill to cut a few tenths bigger.

Figure 6–72 A standard dovetail o-ring.

Figure 6–73 Holding multiple parts securely.

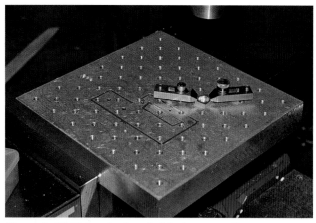

Figure 6–74 This sub-plate holds down small parts.

Figure 6–75 The strap clamps hold the parts.

Figures 6-74 and 6-75 show one of the best things I have ever made. This small 6 × 6 sub-plate with small-sized strap clamps is great for holding down small parts. You can also screw a parallel or a sheetmetal V-plate to the sub-plate as a fence for doing multiple parts.

Tipping machine heads. After tipping the mill head, return to zero with a square first (Figure 6-76). Tram with an indicator after you have the head close using your square. You will get within the small range of the indicator easily with this quick visual method and save yourself some time indicating.

Most machinists hate to tip the head of a milling machine if there is any possible way to avoid it. You see lots of special fixtures and tricks for avoiding taking your well trammed head out of alignment.

On a Bridgeport or similar make of turret milling machine, there is a favorable direction for rotating the head if necessary. There are two possible directions to set the head at an angle. When facing the mill as you would during operation, the four bolts you are looking at unlock the most favorable direction to rotate the head—if you find you must.

This right or left rotation is much easier to re-align when you need to. The reason is the axis of rotation is on the centerline of the head of the machine. For the fore and aft angle adjustment the pivot point is well off the center of the head and makes re-alignment much more difficult. So tip right to left if you have to tip the head.

Figure 6–76 Returning to zero with a square.

Here is a great way to avoid tipping the head in the milling machine. The tilt plate shown in Figures 6-77 and 6-78 can adjust up to 50 degrees of angle relative to the table surface. It's great for those quick angle jobs that are a hassle to tip the head for.

You can set the angle with angle blocks or use can use an electronic level that was zeroed on the table surface (Figure 6-79). If you really need to you can even do compound angles with this setup.

When setting precision head tipping angles, you can indicate the face of a correctly-set sine bar for those odd angles (Figure 6-80). You can also use a toolmaker's ball or large ball bearing to accurately align the axis of the spindle with a rotary table axis once the head is tipped (Figure 6-81). These two tricks will come in handy in the next section.

Figure 6–77 This tilt plate adjusts up to 50 degrees.

Figure 6–78 The tilt plate is good for quick angle jobs.

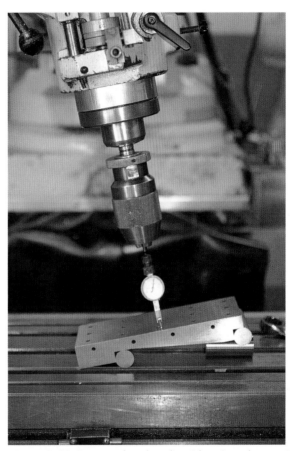

Figure 6–80 Set your head angle with a sine plate.

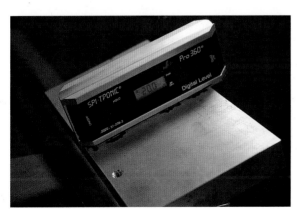

Figure 6–79 Using an electronic level.

Figure 6–81 Aligning the spindle axis using a ball.

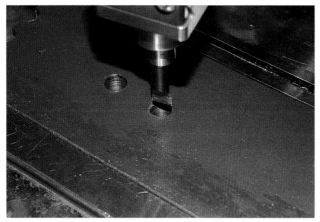

Figure 6–82 Establishing the true centerline of the hole.

Figure 6–83 Drilling and following with a plunge.

For accurate true position location, drill undersized and then use a single point boring tool to establish the true centerline of the hole (Figure 6-82). After a light cleanup cut with the boring head, you can finish ream the hole to size or just use the boring head. Alternatively, you can drill and follow with a plunge with an end mill (Figure 6-83).

Figure 6-84 shows a different twist on parallel retainers made from sheetmetal. These leave the area under the part clear for who knows what to fall into.

Figure 6–84 Parallel retainers made from sheet metal.

6.3 Spherical Surface Generation in the Manual Milling Machine

Figure 6-85 shows a unique manual mill method for generating geometrically true spherical surfaces. This technique can be used to machine convex and concave spherical surfaces. The only tools needed other than the milling machine are a boring head and a rotary table, two common mill accessories. If you have a CNC lathe or mill, this is really just an academic exercise. The principle is interesting in that it is self correcting and self proving, which is not true with CNC equipment. If you don't have any CNC equipment, you can add a neat trick to your toolbox.

Figure 6–85 Generating geometrically true spherical surfaces.

Figure 6–86 Working with convex surfaces.

Figure 6–87 The cutting edge faces inward.

I learned this years ago from my old toolmaker friend Charlie. It's one of those old-timer tricks that I have not seen used anywhere before. When he first told me about it, I was skeptical until I tried it. If you have a computer drafting program, you can make short work of the math and set-up angle. This method is far superior for forming tools and beats the pants off the swinging arc fixtures because the spherical surface is a true geometric generation. The spherical form is limited only by the accuracy of the machine spindle and the rotary table—two intersecting circular paths that produce a true spherical surface.

Imagine a cutting tool that only cuts a hollow circle, kind of like a hole saw. When the cutting tool is set at an angle other than the axis of the rotary table, and the part is rotated under the cutting tool, a spherical surface is generated.

The head is tipped at an angle that represents the chord of the desired spherical segment. A single point cutting tool is used and, depending on whether the form is concave or convex, the cutting edge is reversed. For convex surfaces, the cutting edge faces inward as shown in Figures 6-86 and 6-87. For concave surfaces, the cutting edge faces outward as it would in normal boring head work.

As the cutting tool is advanced into the work, the rotary table is rotated through 360 degrees. In this case, the rotary table is also fed into the tool along the *x*-axis (Figures 6-88 and 6-89). There is an important relationship between the angle of the head and the diameter that the boring head is set.

Figure 6–88 The rotary table is rotated through 360 degrees.

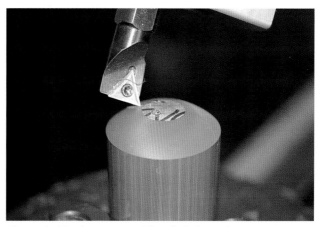

Figure 6–89 The rotary table is fed along the *x*-axis.

Figure 6–90 The black ring is the cutting edge path.

Figure 6–91 The rings are in full contact with the surface.

When you first try this method, I suggest you use plastic so you can quickly see exactly what is happening before you try this on important parts. Don't experiment with hard-to-machine materials when learning, unless you really like turning the rotary table crank. There are three parts you must understand to get controllable results. The first involves the basic calculations. The second is the setup and the third is the execution—actually doing it.

A single-point cutting tool sweeps through a circle that has no thickness on one side of the cutting edge. If you think about how a ring of any size smaller than the spherical surface can lay in full contact with the sphere, you can visualize how the cutting action takes place.

The material that projects into the ring is cut away as the part rotates under the cutting tool. This leaves a spherical surface the size of the ring. The black ring is the actual path of the cutting edge (Figure 6-90). You can see how all the rings are in full contact with the surface no matter if the surface is convex or concave (Figure 6-91). Any plane that cuts though a sphere produces a true circle no matter what the angle.

In Figure 6-92, we can see the basic graphical setup for cutting a full hemisphere of two inches in diameter. The chord in this case is 1.414 inches. This is the diameter the boring head would be set at (1.414) or a little larger to cut this diameter.

The spindle would be tilted at 45 degrees relative to the rotary table axis to cut a full

hemisphere. You can see by inspecting the drawing that no other angle will produce a full half-sphere. The axis of the spindle must be perpendicular to the chord of the segment. The spindle centerline is the midpoint of the chord. The chord is also the hypotenuse of the right triangle that is the maximum rise of the radius and the distance from the centerline to the endpoint of the arc.

For other radii and partial segments, a little math is required to get the chord and the angle. We can use our drawing example to illustrate the math. There is no official name for the diameter that the boring head is set to, so I call it the "Swept Diameter" for our examples, shortened to SD. For OD work, the swept diameter should be set at the chord size or larger. For ID work, the head should be set smaller or the same as the chord.

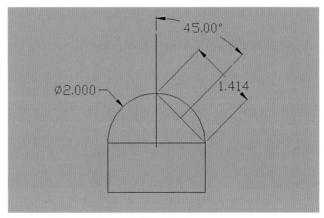

Figure 6–92 Cutting a full hemisphere.

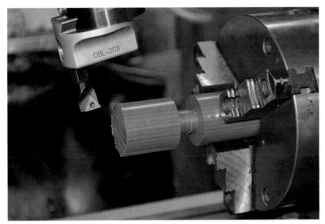

Figure 6–93 Using a .75 diameter stem.

Figure 6–94 Pre-necking the blank.

Angles less than 45 degrees produce less than a full hemisphere. Angles over 45 degrees produce greater and greater portions of the sphere until you reach a maximum of 180 degrees for a full sphere. Once you go past 45 degrees, the boring head must be set accurately to the chord length before you reach the finish diameter. You can adjust this as you rough the part, taking measurements as you go.

We can't in actual practice cut a full sphere in one setup. We still have to hold on to the part and rotate it somehow. In this example, I decided that a .75 diameter stem would be enough to hold on to (Figure 6-93). I pre-necked the blank so the cutting tool had clearance (Figure 6-94). To produce a full sphere, you must use two separate holding setups. In Figures 6-95 and 6-96, the spherical section in complete.

Figure 6-97 shows the drawing I used to set up the previous example. The head angle was set at 11.01 degrees using the sine bar method in the previous section. I started out with the boring head, cut a diameter larger than 1.963 inches, then adjusted the boring head as I cut because I was able to take direct diameter measurements off the part.

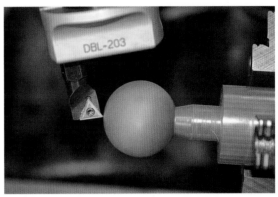

Figure 6–95 Producing nearly a full sphere.

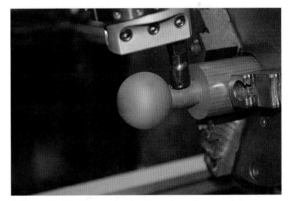

Figure 6–96 The spherical section is complete.

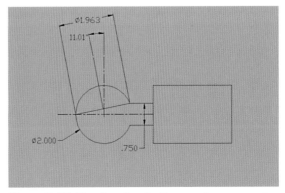

Figure 6–97 The drawing for the previous example.

Figure 6–98 A concave radius.

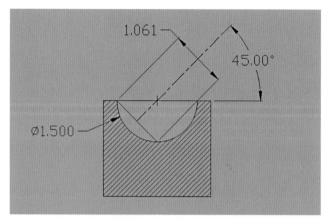

Figure 6–99 Determining the boring head setting.

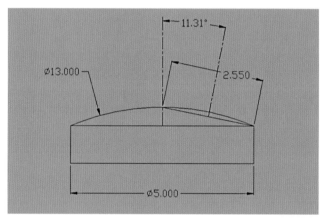

Figure 6–100 A large spherical segment.

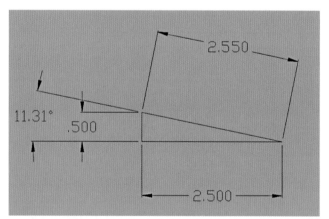

Figure 6–101 Solving for the right triangle.

Figure 6-98 provides an example of a concave radius. The tool edge is pointed inward just as it would be for normal boring work. The swept diameter in this example is the chord 1.061. The head angle would be set to 45 degrees. This produces a full hemispherical cavity.

To calculate other radii and segments, here are the key factors.

$$SD = (D) \, Sin \, \theta$$

where

SD = Swept Diameter (boring head setting)

D = Desired spherical diameter of part

Sin θ = Sine of the spindle angle

To determine the boring head setting, SD = (D) Sin θ. In the example in Figure 6-99, our calculation would look like this,

$$(1.50) \,.7071 = 1.061$$

For segments other than full hemispheres, you will have to calculate the chord of the specific desired spherical segment. Note that the head angle setting is important for getting good results. Figure 6-100 shows another example of a large spherical segment.

We are solving for the right triangle that is the rise of the segment and the hypotenuse to the arc endpoint (Figures 6-100 and 6-101). Take a look in the shop math section for a description on calculating the chord length and radii of arcs.

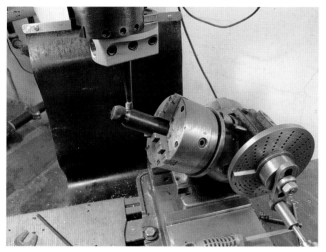

Figure 6–102 Rounding a round.

You can also use this trick to cut large radii on the corner of rounds that would be difficult or impossible with form tools (see Figure 6-102). The only requirement for accuracy is that both axes are co-planar in at least one plane. It's best to start the boring head a little large for OD work and a little small for ID work to keep you on the maximum material condition. The boring head can be adjusted on the fly or ahead of time.

When I have set the boring head ahead of time, I use a height gage on the surface plate or do a short test cut to adjust the tool precisely. When actually cutting, advance the cutting tool from the outside toward the center of rotation. It makes it easier to see when you are exactly on center without re-cutting the entire surface.

In the next chapter, I will discuss the CNC mill and some of the complex things that can be achieved. In order to fully appreciate what the CNC mill can do, it was necessary to have this chapter on the manual mill directly preceding.

In the age of computer-controlled mills, many young machinists may not have had a chance to operate a manual mill for any length of time. Hopefully some of the things I have illustrated here will give the reader some appreciation what us old timers had to go through to get things done before computers.

CNC Mill

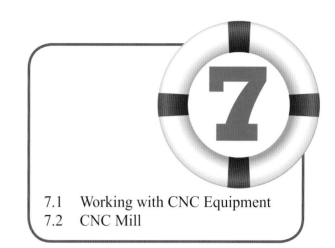

7.1 Working with CNC Equipment
7.2 CNC Mill

The earliest claimed numerically-controlled machine that I can find was a milling machine unveiled in 1952 at MIT. Much of the early research was sponsored by the military as a technology investment in the future. Like many things, the tools and requirements of the military help push the frontiers of technology by investing large sums of money in areas that may have no immediate use in industry, unlike the structure of private research where the financial payback is on a much shorter leash.

These machines differ from their modern counterparts in that they did not operate directly by computer control, but rather by commands calculated by a computer and read by the machine through a punched tape. Most of these early NC machines were converted from existing machinery as opposed to built from the ground up with the intention of automatic control.

7.1 Working with CNC Equipment

Anybody with half a brain can immediately see the usefulness of computer-controlled equipment. The ability of a machine to telescope the work processes so they can run in parallel instead of series can turn a single machinist into a one-person army. Think of it this way: a manual machinist does one man-day's worth of work in one day, right? What if they could do four or six man days of work in that same day? Would you be interested in charging out your hourly shop rate multiple times a day for every machinist? I'm willing to bet you are!

These machines should be thought of as new and faster tools that are available to manufacturing; they allow the company to be more competitive or to increase potential profits. They should not be thought of as machines that displace or obsolete manual machine operators. It's just technology, folks! Every industry, every sport, and every army is looking for a technological edge over the competition.

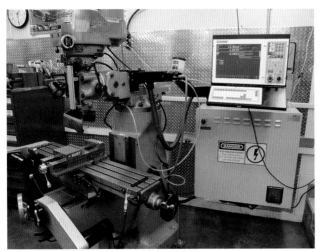

Figure 7–1 Converted manual to CNC mill.

I tell people that I don't really want to be on the cutting edge of technology, but I sure want to be riding on the blade. Like many of the things I have learned over the years, I was fascinated by these machines and their ability to do the tasks they do with such simple looking instructions. At the time it was all secret code, which I didn't understand.

For many machinists the learning curve is steep, especially adding in the complexity of using desktop computers, which now do a large part of the programming duties.

A few words of advice: Don't go the way of the Wisconsin ice cutters! Most experienced non-CNC machinists already know all the hard stuff like speeds and feeds and how to hold things. Most computer jockeys can figure out the controls on a modern CNC machine pretty quickly. The only problem is it takes them ten years to figure out how to hold tools and materials, and then which tool performs what task. The old timers already know that part of the work.

So hang in there; take small, steady steps and, by all means, take advantage of these machines.

One challenge I found difficult when I was learning to run CNC equipment was there was absolutely no sense of feel. Manual machinists develop an accurate sense of feel for something like how hard they can push a tool or how deep a cut they can take with a particular setup. This sense of feel is removed with automatic equipment. I had to learn the actual "numbers" for speeds and feeds when programming CNC machines (see Figure 7-1).

The current crop of CNC machinery is so blindingly fast, it is frightening. For most jobs you cannot even approach the maximum available feed rates for this equipment. The world is waiting for the cutting tool industry to catch up with the machine builders. Currently the cutting tools are the gating factors for most metal removal. There are so many factors that are interconnected with the metal removal rate. Material type, work holding, and operation type are just a few that affect how fast material can be removed. Rarely is the machine speed or power the limiting factor.

My experience with CNC machinery is in the jobbing shop environment. Most of the tips and tricks are related to the highly varied work and many different materials found in the jobbing shop. High-volume production-type work is outside my intended scope for this chapter, but some of the principles and ideas easily transfer to high-volume work.

Try to reduce cycle times for multiple or long-run parts, but never at the expense of consistency or predictability. A machinist or operator can do much more productive and profitable work than babysit a temperamental process.

Work on fixturing and part changing to reduce button-to-button time. This should be the actual measure of a single cycle. If you're just keeping track of how long the machine takes, you're in for a big surprise. On the same note, when you're quoting this type of work, be sure to consider all the factors—not just the programmed cycle time.

Add more comments in the program than you think you need. Memory is not as much of a problem as it used to be. Good programming practice includes comments and specifics about tooling and setup resident in the program (see Figure 7-2).

Try to program in a non-specific machine format. If possible, your programs should be able to run in any equivalently equipped machine. The machine-specific information comes out in the post processor.

Programs that have actually been run and edited at the control are the Holy Grail. Programs downloaded from the machine control should be identified specifically in your program inventory. However your shop keeps track of programs, there should be a simple way of identifying programs that have been run successfully. We add the suffix DL for download when saving a program downloaded from the machine control.

Review the model or electronic information carefully when quoting. There are so many ways the time you have allotted for file conversion or importing can get screwed up. CAM systems work flawlessly on perfect models. Most of the time, the only place to actually get perfect models is at the machine tool shows where they demonstrate the CAM software. Designers find new ways on a daily basis to make the work of the machine shop never boring.

At some point in the completion of the job, the actual part should be compared to the original, unmolested, customer electronic information received when you started. After all the electronic file conversions, model surface cleanup, hole blanking, surface extending, and other steps we do to a customer's model in your CAM system to get it cut, sometimes it will come back and bite you. Go back to the original source.

If possible, build your programs in such a way that a complete part comes off the machine. There is nothing more annoying than having to re-set up because the part did not pass the first article inspection or all your extra spare parts were consumed in setting up the other operations. Parts move and dimensions drift as more material is removed. If you can get a complete part or parts in hand with each cycle, you are doing well.

Figure 7–2 Sample CNC program.

Rather than get specific about types and styles of machines, I'll present this information more generally. Most of what I offer has been learned the hard way, by trial and lots of errors. Each type of job or class of work has optimal machine configurations and setups. I hope you can apply some of these tips and tricks to your specific problems.

Interior slugs. Sometimes it's safer and the machine can run longer unattended if you machine interior slugs into chips (Figures 7-3). It takes a bit more machine time, but can be a real part saver and eliminate babysitting or a bunch of M01 (optional stop) lines.

An alternative is to use a large diameter end mill to remove the slug (Figure 7-4a). The gap between the wall and the slug is large in relation to the slug diameter (Figure 7-4b). This gap allows the slug plenty of room to jiggle around and fall free without wedging. Make sure there are no obstructions

below the part, like a pile of chips to interfere with the slug dropping free.

For big slugs, it's safer to leave a webbing in the bottom and program a M00 (full stop) with an axis retract to give you room to knock out the slug.

Drilling around the perimeter followed by a couple of quick passes with a roughing end mill to within .075 of the bottom was the most efficient way to get this material out (Figure 7-5).

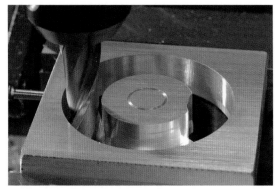

Figure 7–4a Using a large diameter end mill.

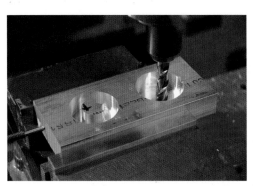

Figure 7–3a Machining interior slugs.

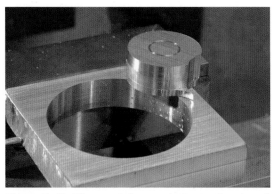

Figure 7–4b The gap is large relative to the slug diameter.

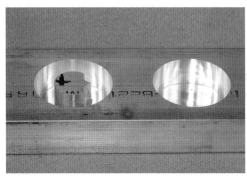

Figure 7–3b Another look at interior slugs.

Figure 7–5 Ways to get out this material.

Figure 7–6 Another look at interior slugs.

Figure 7–7 Another look at interior slugs.

Chip extraction on vertical machines is a big problem. The through hole drilling provided a way for the chips from the rougher to wash out the bottom, and it physically removed a large amount of the material before the milling operation. We knocked the slug out during a programmed stop in a few seconds (Figures 7-6 and 7-7).

Workholding with O-Rings. Here is a neat little holding trick. I think everybody has had trouble holding multiple parts in a single vise to take advantage of the speed of the CNC machine. These soft-jaws were cut to hold the parts with the addition of some pieces of o-ring cord stock (Figure 7-8). The o-ring intrudes into the part pocket by .010, taking up for any size variation in the individual parts (Figure 7-9). Drill the o-ring holes first, then cut the pockets.

Here is another way to use o-rings to help hold parts. Figure 7-10 shows a standard dovetail o-ring groove cut into the movable soft-jaw of the vise. The little bit of o-ring sticking out above the surface grips all the parts securely for a facing operation (Figure 7-11). You can use a piece of Delrin rod instead of the o-ring for holding metallic parts.

Figure 7–8 Using o-ring cord stack.

Figure 7–10 A standard dovetail o-ring groove.

Figure 7–9 The o-ring takes up for size variation.

Figure 7–11 Preparing for a facing operation.

For one or two parts, you can run a second drill cycle after pocketing partially through drilled holes (Figure 7-12). This will save on subsequent de-burring operations. The vertical burrs on the hole edges are difficult to hand de-burr.

If you have the time, you can run around with a smaller end mill and actually radius these edges. For one or two parts, it's hard to justify the additional programming time required.

Profiling. For some kinds of profiling operations you can leave .010 or less in the bottom of the profile to retain the part for the finish pass (Figure 7-13). I like to leave .030 on the bottom

Figure 7–12 Running a second drill cycle.

on the pass just before the finish profile cut. Then on the last pass, I take it down to .010 on the bottom and on size on the profile. The part can then be cut free with a utility knife (Figure 7-14). For this kind of profiling, it is helpful to use bottom-up programming to assure you leave exactly the right amount on the bottom regardless of material variation.

Figure 7–13 Preparing for the finish pass.

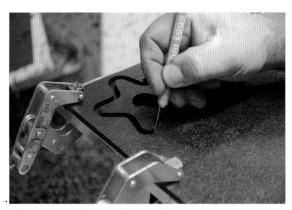

Figure 7–14 Cutting the part free.

Double check your rapid planes when using bolts or clamps to secure a part for full profiling (Figures 7-15 and 7-16). I think everybody has a little box of bolts next to the machine with the heads machined in some very disturbing ways. Count yourself lucky if this is all the damage you can muster up.

Figure 7–15 Bolts can secure a part for profiling.

Figure 7–16 Clamps can also secure a part for profiling.

Figure 7–17 Cut thin material all at once.

Figure 7–18 These parts are .002 thick.

Sandwich thin materials and cut all at once (Figure 7-17). Use a soft cap plate to spread the clamping force over the parts. The parts in Figure 7-18 are .002 thick full hard stainless.

A quick custom-sized sub-plate can make full profiling easier by providing clamp access all the way around the edges. Figures 7-19 and 7-20 show examples I have made over the years.

Figure 7–19 A custom-sized sub-plate for full profiling.

Figure 7–20 This sub-plate provides clamp access.

For parts without interior holes to hold them down, you can use a technique I call musical clamps (Figure 7-21). This requires operator intervention at critical times to move clamps and holding fixtures. It helps if your CNC control will allow manual spindle stop with feed hold from the operator station.

Figure 7–21 Musical clamps.

The height and generous overhang of the profiling plates in Figure 7-22 make clamping a cinch. Treat these as consumable tools to get your jobs out the door. Be sure to sink the screws well below the surface so you can take a light skim cut on the top each time you set it up.

The round base allows me to use the handy plates in Figure 7-23 in the lathe as well as the mill. The V-block gives me secure three-point contact clamping.

For a quick setup and secure holding, I sometimes use heavy pattern C-clamps (Figure 7-24).

Figure 7–22 These profiling plates are easy to clamp.

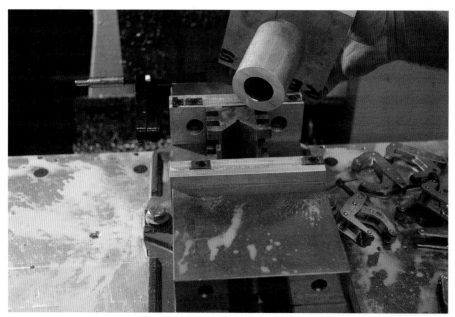

Figure 7–23 These plates can be used in the lathe and the mill.

These heavy duty clamps are strong enough to pick up the milling machine so they should hold your work piece without breaking a sweat. This was a quick profiling job in five or six plates.

I just clamped them to my aluminum sub-plate that is held by the machine vise and I was off and running. Setup time is all of 2 minutes. I even used a little cantilever clamp as a stock stop.

Be careful when stack cutting. Better to make holes fully into chips than risk a loose slug upsetting your whole job. Use a combination of DOC and a number of passes that leaves a thin web between each of the stack layers. You can also pre-drill out most of the material to make the remaining slugs more flexible.

Figure 7-24 Heavy pattern C-clamps.

Soft jaws. Cut a shallow recess into the backside of your soft jaws you intend to keep for the future (Figure 7-25). This makes registering the two sides accurately to one another much easier. Another method might be to add a dowel pin hole or vertical slot in the vise jaw. A short dowel pin could be installed into the soft jaws for accurate location.

This requires a modification to your vise, however, and has the potential to be cut into when preparing the soft jaws.

Store bought soft jaws are hard to beat from a price point of view (Figure 7-26). They leave a little to be desired in the way of available configurations and features. All you entrepreneurs—listen and get busy! One thing we have done to make commercial soft jaws more useful is re-cut the counter-bores and use flathead screws for better location repeatability for repeat jobs (Figure 7-27).

Permanently engrave important offset and program information directly into the soft jaws (Figure 7-28). There's nothing like trying to figure this out in six months when the job comes up again. Extra information available at the machine during setup makes for fewer mistakes and faster, more confident setup (Figure 7-29).

Figure 7–25 Cutting a shallow recess.

Figure 7–26 Store-bought soft jaws.

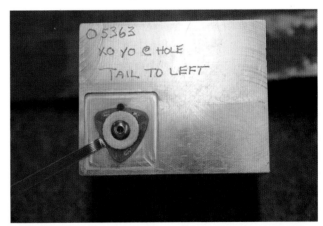

Figure 7–28 Important information should be engraved.

Figure 7–27 Making soft jaws more useful.

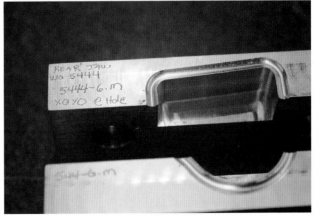

Figure 7–29 Extra information is also helpful.

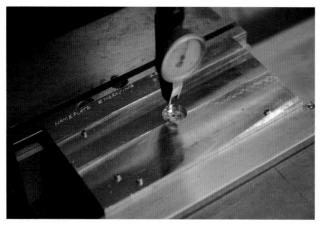

Figure 7–30 Cutting a pick-up feature.

Figure 7–31 Using pick-up features.

Part alignment. Cut a pick-up feature or features directly in your fixtures and soft-jaws (Figure 7-30). There's nothing like a good round hole marked X0 Y0 to set up from the next time you use the fixture.

Figure 7-31 shows an example of using pick-up features so you can easily work outside your machine travels. The dovetail o-ring groove in this large flange weldment was larger in diameter than the Y travel of the machine. It was cut in two halves by flipping the part 180 degrees. An accurate central bore and rectangular feature allowed two super-accurate indicating surfaces to be used when the part was turned for the second half. The triangular plate was attached for this specific purpose. This technique can effectively double your machine envelope.

I've never had to do it, yet, but I heard about somebody doing a job that was just a couple of inches outside the X travel of the machine. The job requirements were such that it really wanted to be done in one setup. The solution was to align the part X-axis diagonally with the machine axis. This put the part axis on the hypotenuse of the machine X, Y travels, and gave the person that extra couple of inches they needed to do the job. With modern CAM systems, this is certainly a simple viable option. On the machine in Figure 7-31, that would be an extra five inches of travel. I don't know about you, but there have been times I would have sold my soul for an extra five inches of travel.

A silver metallic felt pen provides a quick, easy way to retain a tool offset directly on a black oxide tool holder (Figure 7-32). It only takes a second to record an offset and it's cheaper than tags. Heck, I can't even buy the nifty little tags for these BT-35 holders, so I improvised like a true cheapskate.

Figure 7–32 Retaining a tool offset.

Figure 7–33 Methods for engraving.

Figure 7–34 A shallow pocket helps with engraving.

Engraving. Engraving is super sensitive to cutting depth (Figure 7-33). Your engraving will look lousy if your surface is sloped, dished, or not where you think it is. The best approach is to establish an accurately known Z position just prior to the engraving operation. An alternative is to cut a shallow pocket for the engraving to reside in (Figure 7-34). The pocket also helps protect the engraving from damage because it is recessed. By the way, the pocket technique works wonders when you misspell something.

Difficult tapping. You can use an M00 or M01 command just prior to a tapping cycle on difficult materials. Dab a little heavy-duty tapping fluid directly on the tap manually. Water-based coolants leave something to be desired for tough tapping operations. This can make or break a critical job in tough material where a broken tap is not an option. M00 is the preferred method because the default machine condition is a safe one, unlike the optional stop where the operator must remember to have a switch set.

Thread milling. If you haven't tried thread milling, you should. Figure 7-35 shows several different types I use. Sometimes it's the only way to get a thread in a delicate part. Figure 7-36 shows an example of 3/4 NPT in acrylic. With a large tapered thread so close to the edge, it would have been difficult not to break the part with a normal tapered pipe tap. When thread milling, the cutting forces are low and it's easy to adjust the thread pitch diameter for a perfect gage fit. You can also cut a full engagement pipe thread with less than the full depth of hole you would need, even with a short projection pipe tap.

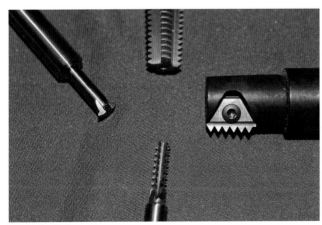

Figure 7–35 Various tools for thread milling.

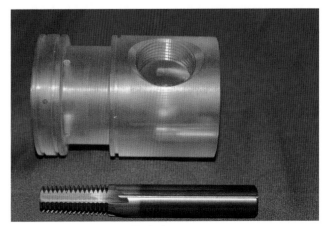

Figure 7–36 Thread Milled 3/4 NPT.

Don't stress out too much about the ramp in and ramp out moves if you are hand programming thread milling. The straight-in approach works fine and is easier to think about the first few times you try it. Most of the time, you are ramping into air or a thread relief. You can use G41/G42 tool compensation if you like to adjust the diameter, but this can be a little tricky sometimes down the hole of an internal thread. I generally do the entire thread without tool compensation and just adjust the X or Y point where the tool starts to orbit for the helix. I put a marker in the program if I am fussing around adjusting this value so I can find it easily. With the single flute-type thread mill, you can also mill those oddball thread pitches that come up every once in a while (Figure 7-37).

Tool runout. I'm sure everybody has experienced the weird phenomena of an end mill or drill lasting for hundreds of parts, replacing the tool "just in case," and then having the new tool only drill or mill one part before failing.

Check a few of your ER-style collet setups for runout if this has happened to you (Figure 7-38). You might be shocked at what you find. These collets are not automatically accurate. If you're having problems with small drill or end mill breakage, check your runout. This is particularly important with small tools that cannot tolerate much runout in proportion to their diameter.

If you find runout, try thoroughly cleaning both the collet and the holder. Then clock the collet in

Figure 7–38 Checking ER-style collet setups.

the holder for minimum runout. The culprit is usually the inside surface of the nut and collets with coolant residue or small burrs on them where the nut touches the collet. Keep these surfaces clean and slippery to minimize problems.

Design recommendations for thread milling. The considerations for the design of thread milled parts are similar to those produced by turning. Space must be left for the threading tool in the form of a thread relief. The relief diameter should be slightly smaller than the minor diameter of the thread, and roughly one full thread pitch in length. The smaller diameter eliminates any incomplete threads by allowing additional cutter clearance. Threads should have at least a full pitch clearance to any shoulders or larger diameters.

Figure 7–37 Using a single flute-type thread mill.

Figure 7–39 An ultrasonic cleaner.

Cleaning and cleanup. A great way to clean your small collets is in an ultrasonic cleaner (Figure 7-39). Small chips and debris worm their way into the collet slits and ruin accuracy. Use a heated solution of degreaser like Simple Green or Omni-All, followed by drying and a dip in a light-weight oil like M-1 or WD-40.

If you are planning your CNC shop, be sure to include a water hose bib near your machines. Machines that operate all day long lose significant coolant water to evaporation. This can be made up with added water or a makeup solution of your water-based coolant to maintain the correct concentration.

Plumb a quick disconnect in line with your coolant line (Figure 7-40). If you make up a short hose with a spray nozzle, used it to wash out the inside of the machine after a job (Figure 7-41). One hose fits all machines. This also makes it easier to steal some coolant for use in another machine in a pinch.

Figure 7–41 This hose can wash out the inside of a machine.

Reducing cycle times. Use your tools for cutting as long as possible to reduce cycle times. You can drill, mill, and chamfer with drill point end mills (Figure 7-42).

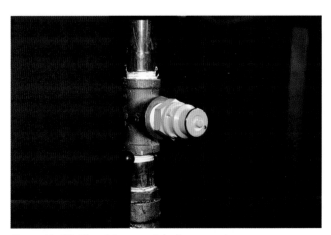

Figure 7–40 Plumbing a quick disconnect.

Figure 7–42 Drill point end mills.

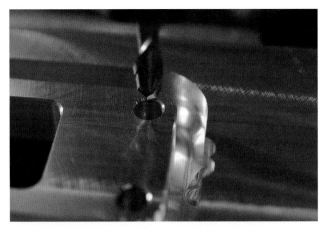

Figure 7–43 Using a tool for several operations.

Sometimes a subtle change in tool selection will allow a tool to be used for several operations, eliminating tool changes and non-cutting time (Figure 7-43). This tip is valid even with modern high-speed tool changes. Vertical CNC machines have less-than-stellar spindle utilization, so anything you can do to keep the tool in the cut is a good thing.

Blank preparation. Consider different methods for blank preparation. The blanks in Figure 7-44 were efficiently rough profiled with waterjet cutting. The material is type 316 stainless. Just because you have a mill doesn't mean you should use it for everything. In this case, the waterjet can fully profile efficiently without any fixturing. Waterjet, plasma, laser, and flame cutting are all

processes you should be familiar with and understand their advantages and weaknesses.

Use every tool, trick, and option at your disposal to get these jobs out the door as quickly as possible. Speed and momentum have a cost saving advantage when applied to part processing.

Figure 7-45 shows the leftover from a job waterjet cut from two-inch-thick stainless; it's so cool I can't bring myself to scrap it out. The parts were then profiled and finished in the CNC mill. Can you imagine roughing out all the material surrounding these profiles with end mills? The waterjet pierced this material with a neat little hole not much bigger than a 1-mm pencil lead.

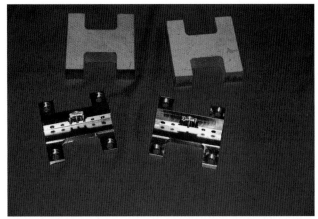

Figure 7–44 Preparing blanks.

Figure 7–45 The leftover from a waterjet cut.

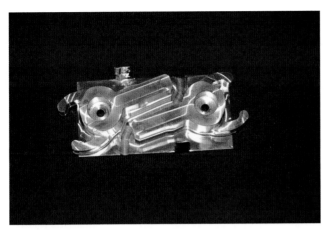

Figure 7–46 Separated ganged parts.

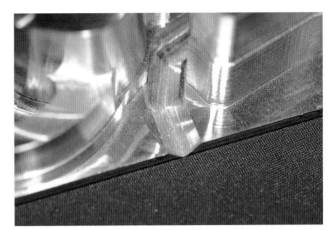

Figure 7–47 Following the contour.

Ganged parts. Here's a trick for separating ganged parts (Figure 7-46). A 90-degree chamfer mill or drill point end mill is used to follow a contour that will be the parting line between multiple ganged parts (Figure 7-47).

Leave .001–.003 at the bottom for an easy break line. These parts can now be snapped off and easily inserted into a pocketed soft jaw for its second side operations (Figure 7-48). You can see the faint raised trail on the backside of the part in Figure 7-49.

Figure 7–48 Creating a break line.

Figure 7–49 The backside shows the break line.

Hole roundness. Circular interpolation is great. However, use a boring head for the roundest possible holes. Single-point boring is still the most accurate for true position and hole roundness. Use single-point boring for precision bearing bores and where critical true position is required, like those found with gear meshes and linear bearing bores.

If you're having problems with hole roundness or true position when circular interpolating, try slowing the feed rate down for the finish pass to even out the quadrant mismatches.

Even with modern high speed controls, this little trick will sometimes give you more consistent results. If your plug gage rocks in one direction when you test the hole, this is a sign your holes are not as round as they could be.

Sometimes you can use this out-of-roundness to your advantage. For sliding fits, a few tenths of out-of-roundness provides an air vent for a blind hole or a place for lubricant to reside.

Logbooks. Keep a machine-dedicated logbook at each of your machines and CAM stations. This logbook gives you a place to record problems and maintenance. It also provides a great place to store quirky bits of information related to each machine—items like, "Check Y-axis limit switch if you get error 1234." If you need to call in outside help, the logbook can be a valuable source of history for the technician. If you record information like metal removal rates for different tool types, these logbooks can become a great training aide. This information can become the company's gold standard of operations, helping at every level from the shop floor to engineering and estimating. On the same note, read the logbooks occasionally and see what kinds of information are being recorded there. If it's worth writing down, then it's probably worth reading.

Document and laminate company setup and operation procedures for each machine. This will pay big returns when training new operators. Keep the length to one operation or task per sheet, then title the sheets. Some examples include *Setting tool offsets, Re-starting program at a particular line, Adjusting wear offsets,* and *RS232 communications.*

You can attach a short wire and alligator clip to your electronic Z height gage so you can use it on non-conductive materials (Figure 7-50).

Fixture parts. Super glue can be used to fixture parts for machining. The glue used in Figure 7-51, made by Loctite, is called Black Max. This interrupted facing cut is a good illustration of its holding power (Figure 7-52). A quick rap with a screwdriver handle removes the part from the mandrel (Figure 7-53).

Figure 7–51 Using super glue to fixture parts.

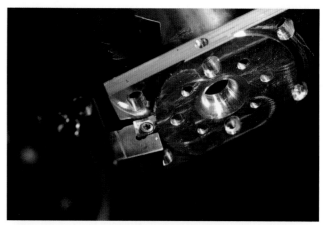

Figure 7–52 The holding power of the glue Black Max.

Figure 7–50 The backside shows the break line.

Figure 7–53 Removing the part from the mandrel.

Figure 7–54 Working with grab stock.

Watch out for excessive heat generation while machining with this method. If the part gets too warm, it will remove itself from the machining area. If you're careful, the plate can be used many times. The replaceable support pegs are a whisker below the top surface of the plate (−.015). Your part should rest flush on the pegs, slightly embedded in the wax.

Figure 7–55 Holding small flat parts for full profiling.

For a few prototype parts in the CNC mill, sometimes it's easier to have a big chunk of grab-stock. The part in Figure 7-54 was milled in one setup and parted off complete in the manual lathe as a second operation. In this case, adding a generous chunk of grab stock was less painful than figuring how to hold it for the second side, cutting special soft-jaws, and posting multiple programs.

Figure 7-55 shows a special fixture I made to hold small flat parts for full profiling. The dark material in the pocket is a special wax called *Dop Wax*, used by jewelry makers to hold stones for grinding and faceting (Figure 7-56). The entire assembly is placed on a hot plate where the wax melts and wets the blank. After the wax cools, the part can be fully profiled without clamps. It works better than double-sided tape because it's truly coolant proof (Figure 7-57). After completion and a little warming, the part can be removed from the wax.

Figure 7–56 *Dop Wax* is used by jewelry makers.

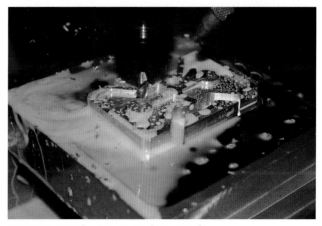

Figure 7–57 The wax is coolant proof.

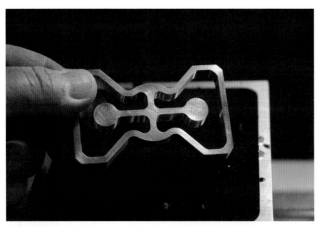

Figure 7–58 Residue can be removed with isopropyl alcohol.

An alternate method is to use Dop wax on a flat, clean plate. After profiling, the residual cold wax can be broken off the back side and re-used. If you have trouble removing the wax, put the part in the freezer for a few minutes; the wax will pop off even more easily. Any residue can be fully removed with isopropyl alcohol (Figure 7-58).

You never seem to have as many tool totes as you would like. I think the available totes for CNC tool holders are overpriced, which makes stocking up on them painful for the small shop. We made the ones in Figure 7-59 in our sheet metal shop from 1/8 aluminum. They are more compact than the commercial units and dirt cheap to make.

This allows you to "kit" a job's tools ahead of time or even store the exact tools for a repeat job efficiently. I even haul these into the CAM area when I'm programming so I make sure I have everything thought out correctly.

Dovetail blanks. For heavy roughing and ripping, use a dovetailed blank (Figure 7-60). This allows you to use a minimum of grab stock with the highest possible security.

The dovetail is a standard 60-degree angle and is .200 in height. Blank preparation is quick and easy. No loose parallels are needed and minimal vise pressure is required for secure holding. You can almost forget to tighten the vise and still hold the blank (Figure 7-61).

Figure 7–60 Using a dovetailed blank.

Figure 7–59 A homemade tool tote.

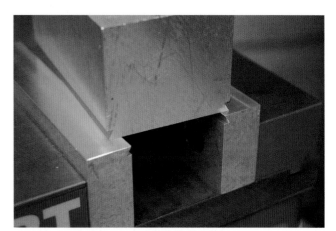

Figure 7–61 Holding the blank.

Grippers. These edge grippers, called Talon Grips (Figure 7-62), are a recent addition to my bag of tools. They are heat-treated edge clamps .060 high. They actually bite into the edge of the blank holding only .060 material.

Figure 7–62 Talon Grips.

In my semi-abusive testing so far, they have managed to hold everything I've asked.

They even come with a nice low profile stop (Figure 7-63). The double-edged grippers are supported deep in a milled groove in the soft jaw and held down with a #10 screw (Figure 7-64).

This is a great alternative to dovetail prepping mill blanks. No prep is necessary and they hold like Barbie with a Kung-Fu grip.

De-burring. Try to de-burr and finish in the machine if possible (Figure 7-65). The added cycle time can be subtracted from the attended machine time.

Use small chamfers on unspecified edge breaks whenever possible (Figure 7-66). 45-degree chamfers are more tolerant of part geometry errors than an equal-sized edge break using a radius.

Figure 7–63 The grippers come with a low profile stop.

Figure 7–65 De-burring and finishing in the machine.

Figure 7–64 The grippers are held down with a #10 screw.

Figure 7–66 Using small chamfers.

Figure 7–67 Double-stick tape works for some applications.

Figure 7–69 Quick change mini-pallets.

Figure 7–68 Sealing out the coolant.

Figure 7–70 Applying a C-clamp.

Part holding. Double-stick tape works well for some kinds of part holding. Only experience will teach you which kinds of parts can be successfully held this way (Figure 7-67). We use a double-sided tape called "Permacel." It seems to be more coolant resistant than some of the other types we have tried. For a little extra resistance to the coolant, we run a bead around of hot melt glue from one of those cheap hot glue guns you see at the hardware store; this seems to seal out the coolant just long enough to get the job done (Figure 7-68).

Prepare your double-stick base plates carefully for best results. I like to dovetail the base plates and surface them all at the same time for an accurate Z position like a set of quick change mini pallets (Figure 7-69).

Clean the plates with IPA before applying the tape. Position the blank and then go around the blank perimeter with a C-clamp (Figure 7-70) or on the arbor press to seat it firmly into the tape. It's important to press the blank down into the tape.

> The holding power of double-side tape is all about surface area in contact. You can prepare the blanks offline while the first ones are cutting if you have several base plates. I always like to use fresh tape for each part just to give me the best chance of success. Or depending on how you look at things, a smaller chance of failure....

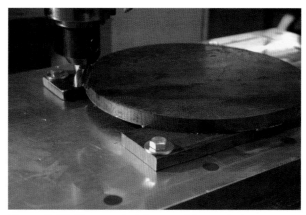

Figure 7–71 Welding the part to the actual hold downs.

Figure 7–72 No extra holding stock is needed.

With certain profiling, you can weld your part to the actual hold downs without the need for extra holding stock (Figures 7-71 and 7-72). Once the profiling is complete, remove the hold down bars or tabs and surface the second side. This keeps the blank thickness close to the finish size, eliminating large amounts of holding stock that make surfacing the second side an effort.

This method works well for open shapes like "C" and "H" that would be distorted from heavy clamping pressure. Welding the tabs from the backside also gives us a little chip clearance and drill penetration room under the part (Figure 7-73).

Figure 7-74 shows another way to use dovetails for part holding. These 440C stainless plates did not have much extra on the thickness for holding.

Figure 7–73 Welding tabs from the backside.

Figure 7–74 Using dovetails for part holding.

I wanted to profile and do all the first side stuff in one setup, so we welded a dovetailed plate to the backside of the blank. We used Everdur, which is a low temperature silicon bronze TIG welding rod to attach the plates (Figure 7-75). These were removed after the first side was machined. The low-temperature process did not warp or affect the 440C plate, and the blank was super-secure while flinging smoking blue chips at 500 SFPM with an inserted death mill. The hole in the center was so we could get a little weld in the center of the dovetailed plate.

Figure 7–75 Attaching the plates with Everdur.

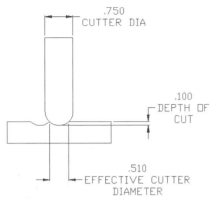

Figure 7–76 Calculating effective cutting diameter.

Figure 7–77 Working with a ball end mill.

All my CNC mills lack rigid tapping, which makes controlling tapping depth a challenge. Inserting a longer pin internally modified the normal tension and compression holders so they have limited compression length. This cuts compression travel to a third of stock. This approach, combined with a modest spindle RPM, gives us good tap depth accuracy. I like the 10 IPM rule: take any tap pitch at 10 IPM; multiply the pitch per inch by 10 to get the RPM. For example, 10-32 tap at 10 IPM gives us $32 \times 10 = 320$ RPM.

Even if you have rigid tapping, look at a reversing tapping head for your CNC mill. They can instantly reverse, whereas in rigid tapping the spindle has to slow and actually stop. If you have huge numbers of tapped holes, this approach can reduce cycle time. Torque control is usually adjustable with tapping head, unlike rigid tapping holders.

Contouring. A trick for three-axis contouring with ball end mills is calculating the effective cutting diameter, then increasing rpm and feedrate to take advantage of this offset (Figure 7-76).

The full diameter of a ball end mill is rarely fully engaged in fine step profiling and surfacing (Figure 7-77). If the actual end mill diameter is used to calculate the speed, you are likely running 50% too slow.

A buddy taught me to use hot melt glue to support a part for second operations (Figure 7-78). In this example, the floor was very thin. When the back side was surfaced over the thin floor area, some chatter produced an unacceptable finish. By filling and supporting the thin floor with the hot glue, the second side came out flat with an excellent finish. Thin webs and floors and overhangs can be supported with this cheap easy material. On most smooth machined surfaces, if can be plucked off without leaving residue (Figure 7-79). Use a little spray mold release to make the removal even easier. Denatured alcohol can also be used effectively to help separate the solidified glue from delicate features.

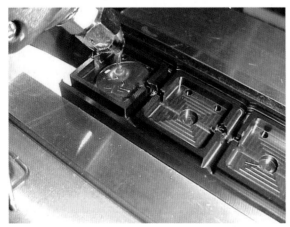

Figure 7–78 Using hot melt glue.

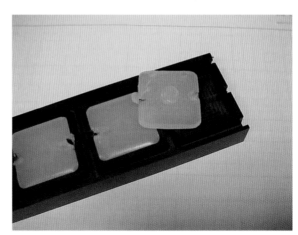

Figure 7–79 Removing the glue.

Figure 7–80 Supporting an unholdable part.

Figure 7–81 Facing and contouring the first side.

Another trick gleaned from my friend is this interesting use of semi-crash proof soft metal or plastic bed plates on the milling machines. At my friend's shop, they prepare the part blanks and secure them to the bed plate with Permacel tape. Plastic locating pins are used to index the part for the second side operations. This technique, used with hot melt glue filling, allows the un-machinable to be machined with ease. The next set of figures show this technique in action.

In this example, hot melt glue supports an unholdable part for two-side machining. This example is held with Permacel double-side tape to a freshly surfaced sub-plate (Figure 7-80). The first side is faced, then surface contoured to the halfway point of the part (Figure 7-81). Be sure to overlap the parting line a little if you are using a ball end mill.

After finishing the first side, the cavity and part are sprayed with a little mold release and filled with hot melt glue (Figure 7-82). If you have a large volume, filling an inexpensive Teflon-coated pan on a hotplate makes quick work of melting the glue.

Locating pin holes are added at this point so the second side is indexed perfectly. Plastic pins are used in case the contouring tool bumps into the pin while the part is cutting. They are only needed when the part is indexed and secured into the double-backed tape.

I leave them in because it's easier and I'm paranoid when using double-side tape.

After the glue solidifies, re-face the first side to your Z zero point (Figure 7-83). Use low rpm to prevent melting the glue.

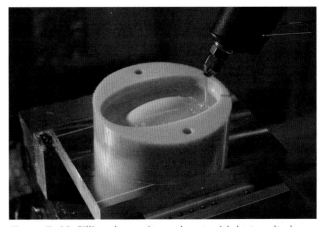

Figure 7–82 Filling the cavity and part with hot melt glue.

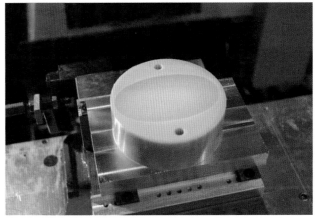

Figure 7–83 Refacing the first side.

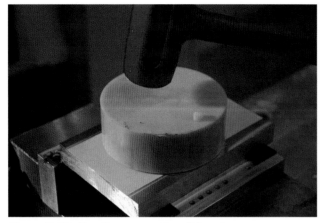

Figure 7–84 Creating a smooth solid surface.

Figure 7–85 Exposing the hot melt glue.

We need a smooth solid surface for the second side so the double-back tape has enough surface area to hold it (Figure 7-84). This method is great for those delicate parts that cannot tolerate any clamping forces without distortion. In Figure 7-85 we see that the surfacing of the second side has exposed the hot melt glue I filled the first side with. Technically the part is floating in space right now, with no connection to the original blank other than the glue.

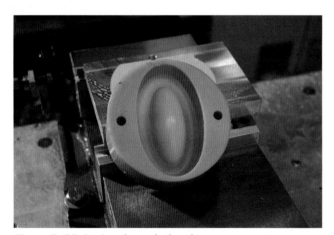

Figure 7–86 Seeing through the glue.

Figure 7–87 A difficult part to hold.

You can see though the hot melt glue a little in Figure 7-86. For the next step, I simply band sawed the excess material to get close to the part.

Be sure to run the saw slowly because it will re-melt the glue and make a mess of the saw. The last hunk of hot melt was pulled off the part, exposing the finished part.

Figure 7-87 shows an example of a three-dimensional part that would be very difficult to hold with any normal commercial fixturing methods. For thin shell or other impossible holding jobs, this method can produce fantastic results.

As available software becomes more powerful and easy to use, the work holding problems become more difficult. Having some nontraditional tricks up your sleeve can allow you to produce some complicated geometry. Another method along the lines of hot glue is to pot the machining blank in automotive body filler. This is a quick drying polyester material that is strong and inexpensive. It can be removed with common solvents like acetone. Obviously whatever material you are machining will also have to be resistant to the removal solvent.

Figure 7–88 A high-speed steel cutting tool.

Figure 7–89 Working with high-speed steel tools.

High-speed cutting. Don't always assume that high-speed steel cutting tools are inefficient on modern CNC equipment (Figures 7-88 and 7-89). On smaller machines, it's typically difficult to take full advantage of the potential of large diameter carbide tooling. On soft materials, like aluminum and plastics, high-speed tools can be run very efficiently at a fraction of the opportunity cost of an equally-sized carbide end mill. For day-in day-out production work on heavy rigid machines, solid carbide or inserted tooling still provide your best tooling cost per finished part cost. We like high–speed, fine-pitch roughing end mills because of the excellent chip control and metal removal rates.

Three-jaw chucks. For round parts that have a small diameter variation, a three-jaw chuck mounted to the table is a pretty handy thing (Figure 7-90). Be sure to make some precision spacers to go between the table and the chuck. The chips are difficult to get out of a chuck mounted flat on the table without spacers. Vise soft jaws cut to accept a diameter would put the true position out by the diameter variation of the part. In this example, we have two diameters with the same end features. Be sure to mount the chuck so you can spin the key 360 degrees.

Clearly mark the machine travels and rated capacities directly on the machine (Figure 7-91). This information seems to come up frequently in the planning meetings and the shop. The more machines you have, the harder it is to remember. In case you're wondering, yes, a few times it's come down to three decimal places.

Off you go now. Good luck and keep the meter pegged.

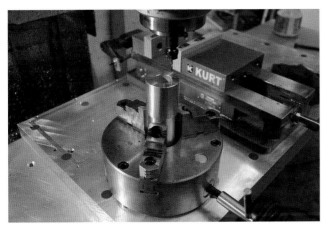

Figure 7–90 The value of a three-jaw chuck.

Figure 7–91 Record important information.

CNC Lathe

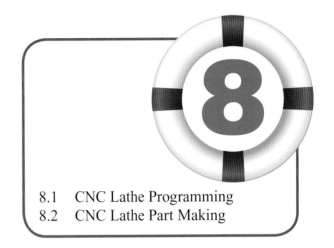

8.1 CNC Lathe Programming
8.2 CNC Lathe Part Making

When I started in the machining world, I started like many people on the manual lathe. I have a special place in my heart for lathes. In my experience, there is a lopsided ratio of CNC mill and CNC lathe machinists. There are about ten times as many mill CNC machinists as there are CNC lathe people. Why is this? I'm not sure, but my theory is that the lathe is more difficult. I will probably get a flurry of letters telling me how this is not true, but still there remains the difference in the number of machinists who are proficient with CNC lathe operations.

I have heard the argument that the CNC lathe only has two axes; therefore, how can it be so hard? After all, the mill has three axes so it seems more complicated at first glance. Like many things, I have some ideas of my own why there are so many more mill machinists than lathe machinists.

CNC lathes are easier to crash than mills.

There have been excellent improvements in what is available for entry level CNC machines over the last decade. Many manufacturers have recognized the need for machines with simplified setup and lower cost points. The explosion of the home and hobby CNC market has driven some of the changes we find in the mainstream machine builders offerings. Today we find a good selection of machines that anybody with the garage space and inclination can now have a full-function CNC shop.

My belief is the simplification of machine setup and the ability to use embedded canned cycles for common operations has brought the CNC machine home—literally. The needs of the smaller job and R/D shops have changed along with the machines. Shorter runs of smaller-sized parts highlight the need for quick agile setup and simplified programming. Functioning as small R/D operations, home-based shops have taken advantage of this class of machine to produce complicated geometry or multiple parts with fast and simple setup in an affordable machine. These home or tool room type machines have been engineered from their foundations to be used in the home, hobby, or small startup environment. One telling feature in this machine type is that most of the machines are offered wired for use with single phase current like you find in residential areas or office suites.

CNC lathes in this class have reverted to the original roots found in the manual lathe. Many of the machines look and function as manual machines as well as semiautomatic high function CNCs. This manual lathe familiarity, combined with easy programming, makes entry into CNC lathe use much less formidable. These aspects, along with lower pricing and options that are easily customizable and upgradable, lower the entry bar enough to make the decision to buy CNC lathe easier than ever. This shift in the market opens the door for small shops to leverage one skilled operator running several machines to increase capacity or speed complex part production.

Most of the CNC lathe programming and setup concepts shown in this chapter are easily transferred to the small shop and hobby environments. The same problems come up regardless of the size or complexity of the machine: programming, setup, work holding, and cutting tools. If you can get a handle on these core skills, then the CNC lathe becomes another tool in your box.

Figure 8–1 CNC lathe.

8.1 CNC Lathe Programming

The typical CNC lathe program is quite short compared to some of the mammoth CAM generated programs seen in milling machines. If you're lucky enough to have a modern lathe with a large memory and a CAM package to program your nice lathe, then some of these programming tips and tricks may not mean much.

Since I am a lathe CAMLESS twit, I have set up my program library into families of parts. I use a couple of generic starting templates that get the machine into a basic configuration to start adding the program elements and events.

Using multiple windows in the text editor, I can copy and paste from one program window to the next (Figure 8-2). Once you have a little inventory of working tested programs in a few materials, it becomes easy to leverage this work against another job. In the world of the low-volume jobbing shop, anything you can do to speed up the non-part-making portion of a process is worth trying.

Copious notes within the program help speed the identification of specific elements that you may want to re-use in another program. This is a great use of the lag time while running parts waiting for the machine. You can make up several generic program scenarios that come up frequently that just need a little quick editing to run.

Clearances. Use a full-scale drawing to check tool clearances with the actual tools you plan on using (Figure 8-3). "When in doubt, check it out!" Down in the bottom of a deep bore is not the time you want to find out the tool didn't have enough clearance, Clarence!

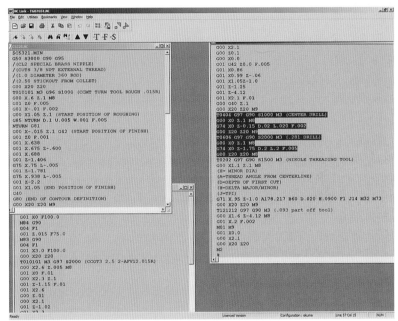

Figure 8–2 Copying and pasting among windows.

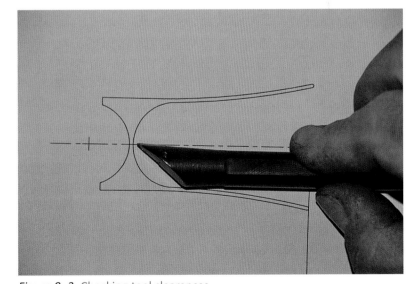

Figure 8–3 Checking tool clearances.

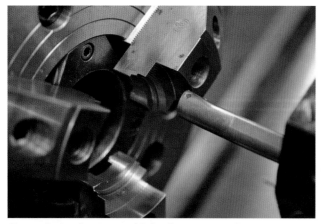

Figure 8–4a Cutting multiple diameters and recesses.

Figure 8–4b Finishing a part in one cycle.

Soft jaws. For many parts, it's really nice to have them come off the machine complete. It's not always possible or practical to do this. One trick for helping is to cut multiple diameters and recesses into your soft-jaws (Figure 8-4a) so you can flip a part around after a pause and finish it in one cycle (Figure 8-4b). There's nothing like having the machine set up to run complete parts when the customer calls to increases the quantity. Programming and setting up to complete parts in one setup is a way of leveraging the initial setup to save future setup time on repeat jobs.

Don't forget about your milling machine when you are creating soft jaws for the lathe (Figure 8-5). Inserts and special pocket jaws can be made easily in the mill. The insert method is an easy way to use standard chucks and collets to hold some really weird shapes.

In Figure 8-6, simple bored soft jaws hold the milled insert for precise eccentric turning (Figure 8-7). It's always nice to complete a part from one side. One thing that keeps you from doing this easily is the final little chamfer or edge break on the inside diameter of a bore that you are parting off into.

Figure 8–5 Creating soft jaws.

Figure 8–6 Using simple bored soft jaws.

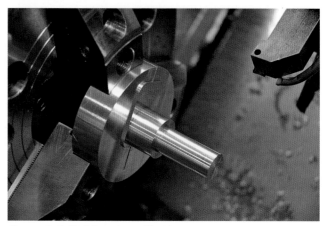

Figure 8–7 Holding the milled insert.

Figure 8–8 Using a threading tool.

Figure 8–9 The tool is modified to 45 degrees.

Pre-camfering. You can use a threading tool modified to 45 degrees to pre-chamfer the ID where the parting tool will break through (Figures 8-8, 8-9, and 8-10). You can profile the chamfer with a standard internal grooving tool. If you choose that route, be sure to pre-groove straight in a few-thousandths deeper than the chamfer before you profile with a regular grooving tool. It can completely eliminate a second operation and the associated handling and price of one extra tool change.

Keep a simple, easy-to-edit program in the control for boring soft jaws or any other repetitive operation that your shop encounters (Figure 8-11). This is a very common setup event in the CNC lathe. If your control supports parametric programming, all the better. A soft jaw setup might be as simple as changing a couple of values at the beginning of the program and touching off the Z axis.

Figure 8–10 Pre-chamfering the ID.

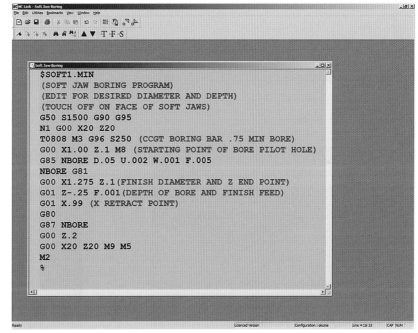

Figure 8–11 Maintain a program template for repetitive operations.

Figure 8–12 A magnetic catcher for delicate parts.

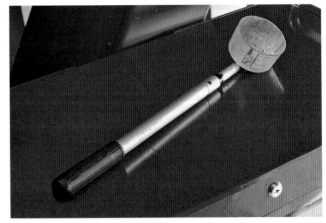

Figure 8–14 A device for catching parts.

8.2 CNC Lathe Part Making

Catching parts. Delicate parts that you don't want going through the chip conveyor can be caught using a magnetic catcher snapped onto the parting station (Figure 8-12). Be sure to put an M01 optional stop or M00 full stop to remove the parts before too many stack up and cause a problem.

If you don't have a parts catcher on your lathe, use CSS (constant cutting speed) when parting right up to where the piece almost comes off. I then switch to a slower constant speed at this point to do the final parting off. This prevents flinging the part and damaging it against the turret or enclosure.

If you really need to catch the part, use an M01 or M00 right before the part separates; use a catch cup on a stick (Figure 8-13). Turn the coolant off in the program so you don't take a bath when you fire it back up.

Don't leave very much to part off with the coolant turned off—just enough to give you time to sneak your catch cup into position will do it. The one I made has a telescoping handle off a rolling tape measure (Figure 8-14).

CSS (constant surface speed). When working with tricky plastics, try turning off the CSS (Figure 8-15). Having control over the cutting speed manually may give you better control of the stringy chips. On some kinds of plastics, chip control in the CNC lathe can be a roadblock to unattended work.

Chip load. You can increase the chip load by slowing the spindle or by increasing the federate to thicken the chip to get it to do what you want. Sometimes, even a simple task like turning off the coolant can help with chip control in situations like this.

Figure 8–13 Catching a part.

Figure 8–15 Turning off the CSS.

Figure 8–16 Special labels for top jaws.

Figure 8–17 Facing the end of a setup bar.

Chuck jaws. I'm sure everybody has had a problem setting the fine serration top jaws in a CNC lathe power chuck. The 1-mm serrations are easy to set one tooth off and screw up the centering on a blank. Three jaws are difficult to measure and set to the middle of the jaw travel. I made some special labels that have the basic sizes engraved on them backward so I can quickly set the jaws to a diameter range and hit all the serrations on the same radius (Figure 8-16). Cost: $5.00 and a little time to engrave on the mill. Be sure to superclean the jaws where these labels stick on. Make it a habit to brush out the fine jaw serrations with a little scratch brush when you re-install your top jaws.

Tool Alignment. Sometimes it's hard to align a boring bar with the machine axis and centerline when the manufacturer doesn't provide flats on the tool. To get around this problem, I face the end of a setup bar and darken it with a Sharpie (Figure 8-17).

Using a turning tool, I scratch a line a tenth or two deep along the turret axis (Figure 8-18). This line is pretty accurately on center and aligned with the turret X-axis angle. It allows me to twist the little boring bar around and have something for lining up the cutting edge. I use a mirror so I can see upside down to align the tool with the scribed centerline (Figure 8-19).

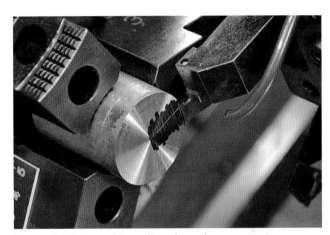

Figure 8–18 Scratching a line along the turret axis.

Figure 8–19 Aligning the tool.

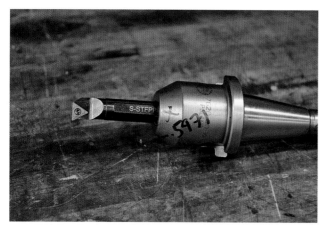

Figure 8–20 Recording the X offset.

Figure 8–21 A back stop screw.

Once you have set up and used a tool, do yourself a favor. When you remove it from the machine, record at least the X offset on the tube or directly on the tool (Figure 8-20). This will speed the tool's setup the next time you use it. The offset won't be perfect, but it can be darn close. If you're wondering, yes that is a picture of a lathe boring bar in an end mill holder. Sorry, you will just have to keep reading to see this idea in action.

Back stop screws. For all my external turning tools, I added a back stop screw that butts against the turret (Figure 8-21). This lets me retain an accurate X offset if I take the holder out of the machine (Figure 8-22). This pays big dividends during the next setup of that tool because I retain the X offset.

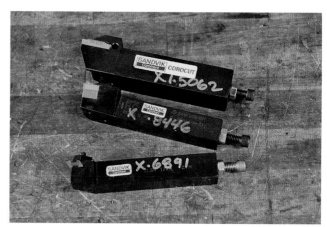

Figure 8–22 Retaining an accurate X offset.

If you have enough machine X travel and your control will allow it, you can run the spindle in reverse and take a light test cut on the OD of a part, with an internal tool to set an accurate X offset (Figure 8-23). You can set the offset without having to make and measure a bore. On my machine, I enter the offset as a negative X value because the tool is into the negative X quadrant. Be sure to check your control to see if it will accept a negative value and still do the offset math correctly. It's almost always easier to measure an outside diameter accurately than a bore or hole.

Figure 8–23 A light test cut on the OD.

Figure 8–24 A simple spindle liner.

Figure 8–25 A few welds finish the liner.

Spindle liners. Shops love to save small disc-shaped parts; you never know when you'll need something like a small flange. Most shops keep a stash of discs that can be repurposed to make simple spindle flanges. For bar pulling operations, it's better to have spindle liners close to your raw stock size.

We made simple spindle liners that drop right in the headstock (Figure 8-24). They use common off-the-shelf pipe and tubing sizes. Have the waterjet or laser cutter zip out some disc blanks next time you have less than a minimum order of cutting. (These technologies are readily available to the home and hobby shop crowd. Software accessibility has improved, making the process available to anyone with a computer.

Toss the discs on the shelf; the next time you need a special liner diameter, half the work is already done. Chop off a piece of tube and weld the discs in place. A couple of quick little welds (Figure 8-25) and you now have an inexpensive spindle liner for that oddball size you need to run.

More cutting. Here's a handy trick for cutting diameters and parts outside your machine's maximum turning diameter. With a custom shop-made tool, we increased turning capacity from 9 inches to over 15 inches. Figure 8-26 shows a close up of the turning tool. We cut the end off a plain old CCGX insert tool holder and welded it to a cold rolled steel shank. Tack weld it first and check the machine movement for clearances and travels before you weld it solidly. This setup puts the tool tip for an OD tool farther back than you can get with normal OD tools.

Using the tool shown in Figure 8-26, we worked well outside our normal turning envelope. The aluminum hexagon in Figure 8-27 was turned using this special tool.

Figure 8–26 Shop-made turning tool.

Figure 8–27 Using the tool to turn this aluminum hexagon.

The Fly

During the summer it gets pretty hot out in the shop. Like many machine shops, ours started out as a warehouse, so air conditioning is a distant want. With sweat running everywhere, I was trying to get a part setup in the CNC lathe. This particular part was giving me some trouble because of a tiny little tool compensation problem. I was hot and frustrated.

To make matters worse, a big, fat, bluebottle fly was buzzing around my sweaty head. Periodically the little pest would land in a drop of sweat that was still connected to my head. With maddening regularity, it would do its little bird bath dance and flee as soon as I tried to swipe at it with a dirty paw. This went on for a while in the heat.

Now I'm a pretty patient person, but this was bordering on a declaration of open hostility. Each time the fly landed, it was like the programming gods were sending reminding me of my flaws as a programmer. I wondered what Hell look like for CNC programmers! My best guess was an endless string of cryptic error messages followed by machine crashes and near misses—with a fly thrown in for good measure.

The next hour felt like an Inquisition conducted by an expert interrogator. Eventually, my feelings went from anger to a kind of resignation. However, I would soon reach the next stage of my downward spiral—I gave up and actually let him creep around on my scalp.

I have experienced moments of almost divine understanding sometimes when I mentally gave up but kept plugging away. This is what happened. I finally figured out the problem with the program. I felt relief at solving the problem and a surge of loathing for my little winged nuisance.

I tested the program and was ready to run it in earnest on real parts. On this part I had set up a bar puller and part counter. It was a small part so I could run over 100 pcs unattended. I opened the enclosure one last time to double check....

The fly flew inside the lathe enclosure.

Blam! The wind from me slamming the door closed probably killed him, but let me tell you folks, it was pure pleasure to push "cycle start" that time. I was not going to open that door unless the collet nose flew off and did it for me. I imagined it was like going through a car wash in a convertible with the top down, except the car wash was spewing a wing wetting mixture of oil, water, and flying shards of metal.

It was that satisfying.

Part Proofing. Figure 8-28 shows a method we use to proof programs and even verify dimensions. We make up several test blanks from MDF board laminated together with glue to get any size blank needed. It's cheap and easy to work with. MDF stands for medium density fiber; it is used as the base for cabinet work using plastic laminate. It is quite accurate in thickness and, more importantly, flatness.

Figure 8–28 Verifying dimensions.

Figure 8–29 Another lathe quick-change tooling idea.

Figure 8–30 Holding an ID grooving tool.

The band sawed laminations are glued together with white carpenter's glue and clamped. These are cheaper and easier to use than huge blocks of wax or expensive butter board. It has no grain so it cuts better than solid wood or plywood.

Quick changes. Figure 8-29 shows another lathe quick-change tooling idea. The commercially-available, quick-change tooling systems are extremely expensive and don't lend themselves to highly varied jobbing shop work. I had several special bushings made that match the taper of cheap and readily-available end mill holders. These bushings have the required accurately-ground tapers and locating keys so the tool indexes accurately. The holder in the picture is an NMTB holder, but almost any taper would work. I like the NMTB because they have a standard coarse thread for the retention end. I just put a bolt in the back and I can switch between tools in seconds rather than minutes or hours.

Figure 8-30 shows an ID grooving tool held in my quick change system using a standard NMTB end-mill tool holder. This quick change setup is generally for ID tools, which typically are more time intensive to set up in the CNC lathe.

Using WD-40. Spray some WD-40 on the inside of the window of your CNC lathe (Figure 8-31). This helps keep the window clear so you can see when you're testing a program. I tried pretty hard to get a good picture of before and after spraying the WD-40 and failed miserably. Anyone who has tried to see through the shower of coolant, desperately trying to see while testing a new program, can appreciate this one. The WD-40 helps keep the window a little clearer while you test the program with your finger poised over the cycle stop button.

Figure 8–31 Using WD-40.

Figure 8–32 Reducing changeover time.

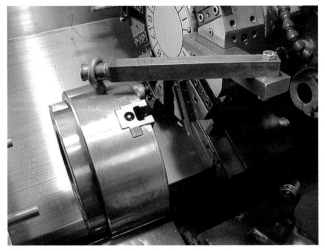

Figure 8–33 Swinging the chuck out the door.

Changing the power chuck on a large CNC lathe can be a real hassle. I made a special tool for helping get the changeover time from chuck to the collet closer down to a half hour instead of an hour or more (Figure 8-32). This pivoting tool clamps in an unused tool position in the turret and the turret is jogged around to help remove or re-position the chuck. Once the chuck is off the hydraulic drawtube, the chuck can be swung out the door (Figure 8-33). In this figure, the chuck is going on the machine. I use two guide pins that fit the retaining bolt hole minor diameter to guide the chuck into position and alignment with the drawtube (Figure 8-34).

In closing this chapter, you will have noticed that many of these tips and tricks are related to setup reduction. For the highly varied work that our shop sees, this is the roadblock to using the CNC lathe for every job. There is a breakeven point with the set up that is dependent on the part and the quantity of customer orders. I would love to run all the lathe work through the CNC lathe, but it is just not possible, at least for our shop and workload.

All I can say is that it is a fantastic tool, but I'm glad it's paid for and does not have to have its spindle turning all day long to pay for itself.

Figure 8–34 Guiding the chuck into position.

The Welding Shop

9.1 Getting Started
9.2 Layout Work
9.3 Some of My Favorite Hand Tools
9.4 Welding Tables
9.5 Brake Bumping

My first experience in the welding shop was when my dad taught me to stick weld in the basement of my childhood house. I was around nine years old and he came home with a brand new welding hood one day. I can remember asking him who the hood was for and being mortified when he told me it was for me because he was going to teach me how to weld.

I donned the nine-hundred-pound, moldering leather welding jacket—complete with the stale sweaty leather smell they all have—and clamped the claustrophobic lightless hood over my head and absolutely loved it. The funny part is that I still use some of the same teaching techniques my dad used on me with the welding wannabes I end up teaching.

I can't remember what my first project was with welding but I know I got into trouble because I went through a 50-lb box of rod just running beads on a flat plate. Something about the arc and the delight of chipping off the slag to expose the gleaming weld underneath caught my interest. The plate started out 1/4 thick and ended up something like six inches thick when I was done. My friends and I used to take turns watching the electric meter speed up when the arc was struck. I wish I could see the electric bill for that month....

The welding shop is where we put it all together (Figure 9-1). Whether you are in a ten thousand square-foot steel shop or in a small corner in your garage stuffed up against the washing machine, the skills required to succeed are the same. In an industrial setting you might be welding the hull of a ship, but even with modest welding equipment you can join those same materials in your own shop to make that log splitter or English wheel you always wanted to build.

Combining the skills of a machinist and welder makes for unlimited possibilities. The tips and tricks presented in this chapter will serve the small shop or hobbyist and the professional metal fabricator equally well. If it's metal and it's cut, bent, burned, or punched, it happens in the welding shop. I find pure welding to be a little boring, so I prefer to mix it up some with layout, forming, cutting,

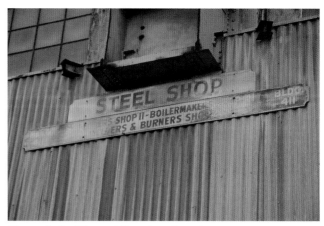

Figure 9–1a The welding shop from large...

and distortion control to keep the work interesting. If you can become versatile in these areas, you can build anything your mind can imagine.

The first step in any fabrication job is to do a little planning. This can be as simple as making a list of materials you will need for a given job. This is not wasted work. Many projects have a way of working themselves into a corner which could easily have been prevented by some up front planning. If it's a big job break it up into manageable bites and best guesses.

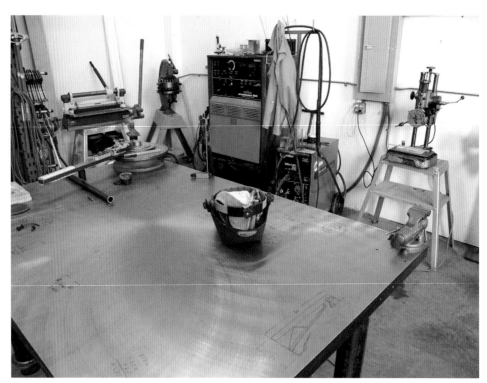

Figure 9–1b ...to small.

9.1 Getting Started

When I start a job in my own home shop, it typically begins with at least a simple sketch. This is the first important planning step in almost any project. From there it depends on the complexity of the project. A simple bracket or tool might only have a single simple sketch needed to finish the job. On something more complicated like a log splitter or race car frame, there might be many sketches needed and other people to consult for the necessary information, materials, and components. There are hundreds of decisions to make to complete a project of medium complexity. Incorporate the information and decisions into your sketches as the planning progresses.

My best results in the shop always seem to happen when I have everything I need at my fingertips—a sketch, some material, and a clear idea of what I want the end result to be.

Try thinking in small sub-assemblies on more complex jobs. I find it's almost always easier to build and control a smaller part of the much larger project. Aim small, miss small.

Be realistic about the tolerances that can be achieved with welded assemblies. Don't make the mistake of trying to avoid slow tricky machine work by substituting even more expensive precision fabrication and the inherent problems that go along with it.

Use these general tolerance guidelines for welded assemblies, as shown in Figure 9-2, for achievable and consistent results. Trying to force tighter tolerances into your welded projects is just asking for disappointment.

Weldment Size	Minimal Welding		Moderate Welding Tolerance		Heavy Duty Welding	
	Close	Normal	Close	Normal	Close	Normal
0–12 Inches	+/– .015	+/– .030	+/– .030	+/– .060	+/– .060	+/– .125
12–36 inches	+/– .030	+/– .060	+/– .060	+/– .125	+/– .125	+/– .250
36–96 inches	+/– .060	+/– .125	+/– .125	+/– .250	+/– .250	+/– .500

Figure 9–2 General tolerance guidelines.

9.2 Layout Work

All the lines, arcs, and angles we mark on our work to guide us fall under the heading of *layout work*. Sometimes work is produced directly from the layout and other times it's used as a reference so we don't make a mistake.

It's tricky deciding how much layout work needs to happen for any particular job. Some jobs require zero layout whereas others are so complicated that a large amount is necessary in order to avoid errors.

Center punching. Grind the tips of your center punch to ninety degrees included angle (Figure 9-3). This angle is extremely durable. Use a soft metal hammer like copper or soft steel to strike your center punch (Figure 9-4). This keeps the hammer from accidentally slipping off the head of the center punch and spoiling the work when you give the punch a solid blast. Hard steel on hard steel is like a banana peel on an icy sidewalk. Keep the tip nice and crisp so you can feel it click into the scribe lines.

Figure 9–3 Grinding the tips to the center punch.

Figure 9–4 Striking the center punch.

Dividers. You will need at least two sizes of dividers at a minimum. A set with a 6-inch radius ability will carry you a long way. I like a set in which you can interchange a pencil with one of the legs. This is great for template work and gives you a reason to steal those lame short pencils from the library or the golf course. For larger fabrications and arc layout, trammel points that clamp to common stock sizes are excellent.

Trammel points that can fit a wide range of bar sizes are superior. The eccentric point type shown in Figure 9-5 will also hold a wooden pencil for paper pattern layout. Fine adjustments are made easily by rotating the eccentric point.

Figure 9–5 Trammel points.

Figure 9–6a A crack along the scribe line.

Figure 9–6b A crack along another scribe line.

Deep scribe lines can be a failure point when forming. Figures 9-6a and 9-6b show examples of cracks along the scribe lines. Use a superfine point marker or a pencil if the part will be subject to high stress. High strength materials and tight radius bend will fail along these little stress risers.

Forming heavy plate sometimes requires the edges to be softened and potential crack starting flaws eliminated. In Figure 9-7a, the edges were not softened and the start of a crack is plainly visible. In Figure 9-7b, the edges of the plate were sanded to remove the ridges left from flame cutting.

In addition, the corners were rounded slightly leading out to the plate surfaces. It's just good practice on flame or plasma cut edges to always skim over with the sander or grinder to eliminate potential crack formation points and to remove the hard scale layer that can damage forming tooling.

Flip the layout for cutting on the vertical band saw. Don't forget about the limited throat on a band saw. You can flip your layout to the opposite side to keep cutting with the stock on the outside of the machine.

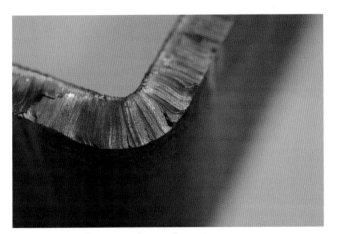

Figure 9–7a The start of a crack appears.

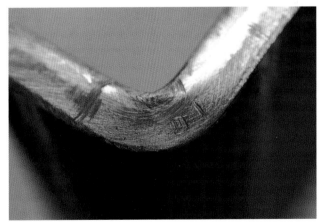

Figure 9–7b The edges were sanded.

Figure 9–8 A simple centering bar.

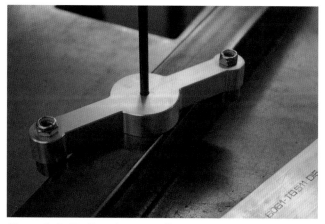

Figure 9–9 Finding the centerline.

A simple centering bar speeds up the all-too-common task of finding the centerline of strips and bars (Figures 9-8 and 9-9). You can use a transfer punch or a carbide scriber made from a broken 1/4-inch carbide end mill to scribe a line. I made mine with little rollers so you can slide it down the length easily. The rollers are spaced equally distant from the center so the center point is always in the center of whatever width bar is used, as long as both rollers are touching.

The center boss is two inches in diameter, which allows you to easily place the center point right where it's needed by burning an inch (Figure 9-10).

Even for a small amount of layout work, a combination square is mandatory. If you have one of those cast aluminum combination squares, do yourself a favor and throw it away right now. Better yet, break in into pieces so nobody pulls it out of the trash can where it belongs!

You can judge peoples' commitment to their profession by the tools they use. If you don't already have one, get yourself a professional forged and hardened combination square set with an additional long blade (Figure 9-11).

These tools last so long that your grandkids will be using them long after you're gone. Normally we use the long blades as straight edges or to set distances off an edge. You have to be a little careful and have a fine touch when using a long blade to check squareness. I like several squares when doing repetitive layouts. All I need to remember is small square small dimension, medium square middle dimension, and big square big dimension. You get the idea.

Figure 9–10 Placing the center point.

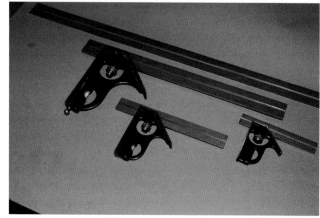

Figure 9–11 Combination square set.

Figure 9–12 Using a centering head.

Figure 9–13 Marking the correct side.

Use a centering head to transfer a line ninety degrees around a corner (Figure 9-12). If you're placing a cross member in relation to your layout mark, be sure to indicate which side of the line the part sits. I mark the correct side with a big X (Figure 9-13). There is not much else worse than having to cut out a fully welded cross-member that's in the wrong place.

Watch out for large tube radii when using your combination square. When you get up to about a quarter-inch wall thickness, a standard combination square is not reliable because it's sitting on a curve (Figure 9-14).

Figure 9–14 Large tube radii.

Figure 9–15 A face extender.

Figure 9-15 shows a face extender I made for working with large tubes. It gives me a wider face for those large tube corners (Figures 9-16 and 9-17). It clamps to my standard combination square head with pointed set screws that just catch the lip of the square and push the two faces together. Make it out of something durable so that, when you slide it along the sides of tubes, it doesn't get knarfed up.

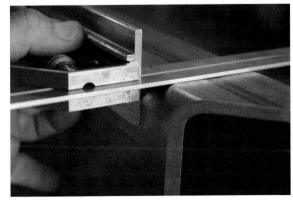

Figure 9–16 Using the face extender.

Figure 9–17 Working with large tube corners.

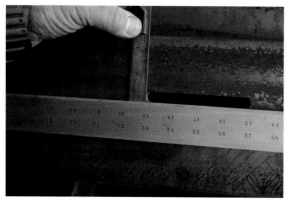

Figure 9–18 Using rules.

Figure 9–19 Combining rulers and calipers.

Rules are generally superior to tape measures for accuracy (Figure 9-18). Quality rulers can be combined with calipers to take large measurements with less uncertainly (Figure 9-19). The overall length of a precision ruler is quite accurate.

Flexible rules come in handy when you need to measure along a curved surface (Figure 9-20) or sneak into a weird spot (Figure 9-21).

Figure 9–20 Measuring alone a curved surface.

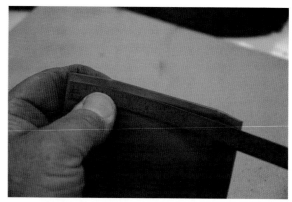

Figure 9–21 Working with flexible rules.

PI tapes can be used with excellent results in the fabrication shop (Figure 9-22). One use is to verify parallelism to a very close tolerance. Alignment of parallel rolls or tubes is a good example (Figure 9-23). Another great use is to verify the actual diameter of an object that is out of round. It is much more accurate than taking fifty measurements and averaging them to get a diameter. This works especially well for rolled sheetmetal tubes and rounds which are less than perfectly round most of the time.

Figure 9–22 PI tapes.

Figure 9–23 Aligning parallel rolls and tubes.

Figure 9–24 Using the proper level.

Figure 9–25 An accurate electronic level.

Your layout table should be properly leveled. This allows you to use a precision level to verify geometric relationships that are difficult to check any other way. A level sensitive to at least .005 per foot should be used (Figure 9-24). Accurate electronic levels are available now that have the ability to set their origin on any surface and compare another surface (Figure 9-25). Angular measurements can be some of the most challenging measurements to take. If your table is skee-wacked, then your work may end up that way also.

You can use the origin reset in Figure 9-26 to compare two surfaces that might be very difficult to check with normal angular measuring tools.

Figure 9–26 Comparing two surfaces.

Figure 9–27 Useful measuring tools.

A simple auto-level or transit can be used to check geometry, or to level large fabrications. These can easily read to .010 over thirty feet. If your datum surface is properly leveled, this is an excellent way to check geometry. This technique is useful for checking features that are not in direct line with one another.

A small bevel is a handy tool for transferring angles to tight spaces (Figure 9-27, third from left). It is a transfer tool. Be careful you don't bump it because there are no marks or divisions on it to verify settings.

Accurate angular measurements are sometimes the most difficult measurements to make. You will need a variety of tools to get the job done, including trammels and protractors (Figure 9-27, first and second from left), and bevels. You will also need a solid understanding of geometry.

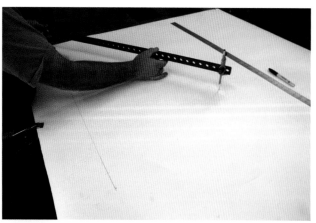

Figure 9–28 Constructing large angles.

Figure 9–29 Using length measurements.

Large angles are sometimes best constructed geometrically using length measurements (Figure 9-28) instead of using a protractor beyond its basic range (Figure 9-29). A fraction of a degree reading error at a large radius means big trouble. You can resolve approximately a 1/4 degree on a good protractor on a good day with youthful eyeballs. At a 24-inch radius, that 1/4 degree becomes plus or minus .10 inches. Be sure to verify your math with a measuring tool that has a scale of some sort on it.

Angular principles, like the easy to remember 3-4-5 right triangle relationship, are an excellent example of a graphical construction. These types of graphical angle constructions are seen in sheet metal pattern work.

It's still easier and more accurate for larger work to use the graphic methods for pattern developments not cut by computer-controlled equipment.

Figure 9-30 shows an example of a typical cutting list. You can see that it has the type and size of each material and any special end conditions. You don't want to be thinking about this stuff when you're at the saw or the shear. I bet you can understand my graphic code for the end conditions of each tube. This thinking should already be done when you step up to the machine. In many of the shops where I worked, there was sometimes a line to use the shear or the saw, so you really needed to get in there and get out as quickly as possible. If you were scratching your head doing any thinking, you got sent to the end of the line. Notice the note for the number of sheets or bars. Be aware of the yield requirements when shearing or saw cutting.

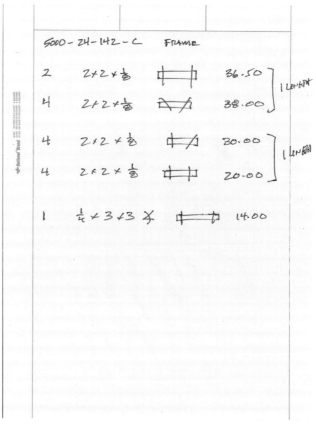

Figure 9–30 A typical cutting list.

Figure 9–31 A freehand burning guide for short cuts.

Figure 9–32 Avoiding interference from molten metal.

Figure 9-31 shows a simple-to-make freehand burning guide for short cuts. The torch tip slides smoothly along the round bar because there is only a small contact point. There's a little clearance under the bar near the torch tip; little bits of molten metal don't get in between the guide and the sliding surface and interfere with the cutting (Figure 9-32). The bar is about an inch in diameter. It's great for quickie little cutting jobs and beats the lame, angle iron-type guides hands down.

The oxy fuel cutting torch is one of the most versatile and portable tools in the welding shop. It provides the ability to heat, cut, and weld with one simple piece of equipment that requires no electricity to operate. When people ask me what is the best type of welding gear to buy first, I always suggest the oxy fuel rig. It goes a long way with a very modest equipment investment. It does require some operator practice to become comfortable and confident.

For longer cuts or where I have to reach across a distance, I made a guide wheel that clamps on the cutting tip (Figures 9-33 and 9-34). The wheel has a 90-degree knife edge so I can make turns easily. A bronze bearing and a shoulder bolt make up the pivot.

Figure 9–33 A guide wheel that clamps on the cutting tip.

Figure 9–34 This guide wheel is for longer cuts.

Figure 9–35 A circle cutting attachment.

Figure 9–36 The attachment is adjustable.

Figures 9-35 and 9-36 show the best circle cutting attachment I have ever used. I almost forgot to mention that I invented it and, no, I don't think it's the best just because I made it. The circle diameter is adjusted by sliding the body along the gas mixing tubes. If more adjustment is needed, you can also adjust the pivot point. With the pivot point shown, you can cut circles as small as 3/4-inch up to a 12-inch diameter circle. Like all circle cutting attachments, it takes a bit of practice to cut a smooth full circle in one pass. It's even harder to cut a smooth circle while trying to take a decent picture of it for a book you're writing…

Figures 9-37 shows a great little tool used to stabilize small parts for precision welding. I call it a welding duck. The peg on the top is used to add weight for a little more pressure (Figure 9-38). The flat bar tail keeps the tool from tipping.

The handy little magnetic TIG torch holder in Figure 9-39 prevents you from laying your torch cable over hot material. It will pay for itself in three months easily by saving on broken torch back caps and melted water lines.

Figure 9–37 A tool for stabilizing small parts.

Figure 9–38 The peg on top provides more pressure.

Figure 9–39 A magnetic TIG torch holder.

Figure 9–40 Grooves wrapping around the electrode.

Don't worry about the grooves in the grinding wheel used for grinding tungsten electrodes. You can sharpen tungsten faster when the groove wraps around the diameter of the electrode (Figure 9-40). You should have a dedicated tungsten grinder in your shop with a good silicon carbide green wheel on it.

I never liked the belt sander for dressing tungsten electrodes (Figure 9-41). I came from a shop where we used short back caps, so the electrode started out pretty short to fit in the torch.

In most belt sanders, short tungsten electrodes are hard to handle to get a close-to-true vertical scratch pattern. The curvature of a normal bench grinder wheel naturally provides hand clearance while grinding.

Sharpen both ends of your tungsten electrodes and have a few close by in the welding area. This saves dozens of trips to the grinder in a typical week. You can make yourself a cool little block to hold the everyday tungstens (Figure 9-42). Just don't run through the house with this little bed of nails!

Figure 9–41 Belt sander for dressing tungsten electrodes.

Figure 9–42 A block for holding everyday tungsten.

Figure 9–43 Gas lenses.

Figure 9–44 Hanging tungsten into tight corners and recesses.

I got into the habit of using gas lenses almost exclusively (Figure 9-43). The larger cup size makes a great guide when welding sheetmetal seams, let alone the ability to hang the tungsten out to reach into those tight corners or recesses (Figure 9-44).

With the large diameter cup of the gas lens, you can rest the cup directly on the seam and traverse along without even having to rest your hand on the hot metal or any other support (Figure 9-45).

With a gas lens, you can extend your electrode to reach into those tough corners (Figure 9-46) and still maintain your gas shielding around the tungsten.

Figure 9-47 shows a cheap-to-make scriber that will actually cut through the mill scale on hot rolled bar and plate.

Figure 9–45 Resting the cup directly on the seam and traverse.

Most reciprocating saw owners can't stand to throw out a blade, even when it s well beyond resuscitation. So there are generally plenty of dead or dying blades available for making a few scribers. In this regard, most shops have a bucket of wasted reciprocating blades available—so who cares if you lose one?

Figure 9–46 Extending the electrode.

Figure 9–47 A scriber that cuts through the mill scale.

9.3 Some of My Favorite Hand Tools

Tools are fun things to collect, but let's face it; there is only a small handful you really use repeatedly on a daily basis. The tools that fall under this heading are almost an extension of your hands and arms. You know just what you can and cannot do with them. I know within a few foot-pounds how much torque I can put on my favorite pry-bar before the tip snaps off. I think about it like this:

A truck with a big NASA insignia pulls up in front of the shop one day and several official-looking guys with dark-rimmed glasses and crew cuts get out. They tell you that the space station has a problem and you are the only one that can fix it.

Figure 9–49 The best rawhide hammer.

As your head swells upon hearing this, the NASA guys tell you to pack only your most necessary tools for the critical repair mission. You are given a strict weight limit for yourself and your tools. Since your belly hangs over your belt a bit, it cuts down on the number of tools you will be able to take.

You have to decide in less than three minutes which tools you absolutely must have with you or you will miss the launch window and the space station will slowly spiral into the atmosphere and burn up. This is how I figure out which tools are my favorites.

Garland Rawhide Mallets. These are the best rawhide hammers ever made (Figures 9-48 and 9-49). You can't beat one of these hammers for essentially non-marking hitting power.

What gives them the powerful punch is the cast iron head assembly. The faces are available in several material types and are readily replaceable.

My two favorites are the smallest #1 and the big monster #5. You will be surprised at what you can move with one of these hammers. Whether you're repairing a global positioning satellite or knocking the dried "residue" off a sewage auger, this is the best soft faced hammer money can buy.

Figure 9–48 Garland rawhide mallet.

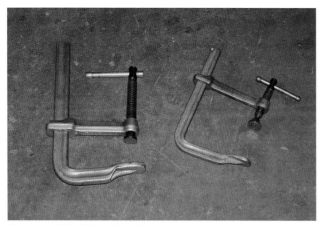

Figure 9–50 Sliding bar clamps.

Bessey sliding bar clamps. The sliding bar clamps shown in Figure 9-50 came on the scene about 30 years ago. The combination of deep throat depth and excellent clamping power makes these clamps a favorite. You can rapidly position them over a wide range and even use them with one hand in a pinch. Try adjusting a C-clamp with one hand occupied. They are virtually indestructible; their only minor shortcoming is they are a little expensive.

The high-quality clamps in Figure 9-51 come in a wide variety of sizes and abilities. In the welding shop, they are worth their weight in gold. I always advocate buying the best tools possible. Even if you do only small amounts of welding, you will find many uses for these quality clamps. I use them on the milling machine with some frequency.

Merit grinder with cutting disc. A small, right-angle pneumatic grinder with a good high quality cutting disc on it covers a myriad of odd jobs in the welding shop (Figure 9-52). Pneumatic tools are more compact than equivalent electric tools. If oiled properly, they will outlive their electron-powered counterparts under heavy duty use. I like the thinner 1/32 thick cutting discs for sheetmetal work and the thicker 1/16th longer lasting discs for weld shop work (Figure 9-53).

The availability and quality of the current crop of electric 4 1/2 inch grinders is hard to ignore. If you do any amounts of grinding, it's a good investment to have more than one grinder setup. If you have one setup for a cutting disc and another with a sanding pad on it, you save hunting down the adapters and are more efficient.

Figure 9–52 Right-angle pneumatic grinder.

Figure 9–51 High-quality clamps.

Figure 9–53 Cutting discs.

9.4 Welding Table

The best material for a welding table is cast iron. The weld berries don't stick to them and when you are sliding steel weldments around on them, they move like they are on bearings. The drawbacks are cost and limited configurations. Keep your eyes peeled for an old planer or large milling machine bed. They are tee slotted for clamps and make a great welding table if you can find one. The only real drawback with cast iron tables is you cannot tack weld temporary fixtures or stops to them. This is a minor inconvenience in my opinion.

Stainless steel on the other hand is just about the worst material you could use for a welding table. Its low thermal conductivity makes the weld spatter, and berries stick without mercy. If you must use stainless steel for specialized work, make it a secondary removable plate that sits on top of your normal welding table.

A steel welding table is the most logical choice for most everybody (Figure 9-54). The table can be readily cut and configured to almost any size and shape. The possibilities are endless.

Figure 9–54 Steel welding table.

Figure 9–55 Flat ground strips.

Install flat ground strips from the machine to the table (Figure 9-55). These are easier to sweep around and cut down on ground cable damage and tripping hazards. If you have expansion grooves in your concrete, these strips can sometimes be inserted edgewise into the expansion to make for a below flush installation.

A Blanchard ground flat surface is best for accurate work. Be sure to stress relieve the plate before grinding to maintain the flatness you grind in. The steel welding table is your basic reference datum. If it's humpy and bumpy, it makes it that much harder to do good work.

Grind both sides. The grinder typically dusts the opposite side anyway to get the plate to sit stable on the machine. As long as you don't specify nutty tolerances and finishes, the price for grinding a large table is not too bad. If one side gets beat up just flip it over and mount the legs on the opposite side.

Check with your local grinder before you order the plate to see how big a plate they can handle if you're thinking about a large table. It's the long corner-to-corner diagonal dimension that matters.

Figure 9–56 Overhangs for clamping.

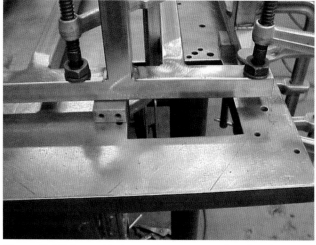

Figure 9–57 Round holes and square notches for clamping.

Clamping. Bolt the legs on so you can re-grind the table surface in the future. Drill the holes all the way through so you can flip the table later on. Provide overhangs along the edges for clamping (Figure 9-56). Put the legs and stiffeners far enough in that they clear your deepest clamp's throat. Putting adjustable feet on the legs allows you to push two tables together and get them to match heights.

When you order the plate for your table, ask for a couple of round holes and square notches in the plate (Figure 9-57). These are very handy for clamping out in the middle (Figure 9-58) as well as dealing with the different types of fabrication that come up. There is no magic recipe for the size or locations; just make sure you have one or two.

Put a couple of tapped holes in the plate in a few places. This helps when lifting or handling the table and can be used for add on fixtures or work stops.

Put a bar between two of the legs to hang your welding clamps on under the table and out of your way (Figure 9-59). Do this on two sides of the table. When you remove clamps, you will need a place to hang them without walking around to the other side.

Figure 9–58 Clamping in the middle.

Figure 9–59 A bar for hanging clamps.

Figure 9–60 Using sawhorses

Figure 9–62 Round legs are easier to attach.

One of the best tables I ever worked on was in a shop that used to be an old gas station. The hydraulic car lifts were still in the floor and the main welding tables were attached to the top. You could raise and lower the table depending on what you were working on. The anti-rotation rod was removed so you could actually spin the entire table with your project on it.

The heavier the work is, the thicker the table. Ours are 1-1/2 thick for serious stability. Generally for extremely heavy work, the tables are much lower than table height, and heavy duty stands or sawhorses are used (Figure 9-60). Be sure to level your stands when you start a big fabrication job. A strip of duct tape on the top of the stand protects your high finish work from scratches.

Weld table features. I always have a vise on my welding table. Make it easily removable if you need the entire table surface. Replace stock jaws with new ones made specifically from copper (Figure 9-61).

Round legs require less welding to attach to a base (Figure 9-62). If you have a large quantity to fabricate, consider using pipe or heavy tube for the columns. The perimeter of a square with the same size is 25% longer than the perimeter of the round. On a big job, that's a lot of extra welding.

Here's a trick I remembered when I shot the legs photos. If you tack weld from inside, the shrinking forces tend to pull the plate and the tube together (Figure 9-63). This example uses the distorting forces to your advantage. This approach only works in limited cases where you have access to the inside of a tube or the tube has a very large diameter.

Figure 9–61 Using copper jaws.

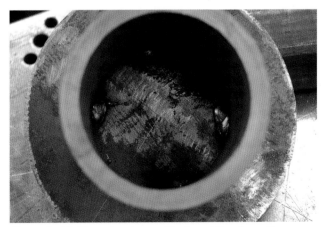

Figure 9–63 The advantage of tack welding from the inside.

Figure 9–64 Flat bars used for gussets.

Gussets. Use standard size flat bars for gussets which you can shear directly from strip (Figure 9-64). It's not necessary to have the gusset go all the way into the corner and it takes more prep to fit them up that way. If you must use triangular gussets, then don't cut them to a sharp point.

Instead, leave a little straight at the end to ease welding around the end. They can even double as lifting points if enough space is left in the corner to pass a sling around the gusset (Figure 9-65).

Figure 9–65 Using triangular gussets.

Arc starting pads. When you have one of those real delicate welding jobs, sometimes it's helpful to use an arc starting pad made from copper (Figure 9-66). This allows the arc to establish itself at a higher current level until you can get your visual bearings without blowing the whole part away (Figure 9-67). Use copper for your pad; I find it works better if you actually touch the pad with the tungsten before you try to initiate an arc. I have no idea why this works—maybe it breaks the oxide film—but the arc starts more easily when you do this.

Figure 9–66 An arc starting pad made from copper.

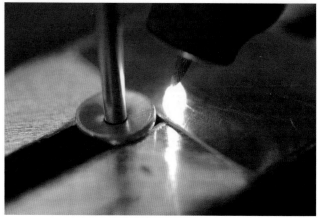

Figure 9–67 Using the arc starting pad.

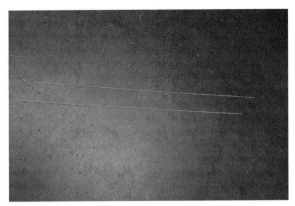

Figure 9–68 The ring test.

Figure 9–69 Imprinting the alloy.

Filler rods. Use the ring test to evaluate unknown aluminum filler rods (Figure 9-68). Most shops have two types on hand: alloys 4043 and 5356. When dropped on the concrete floor, the 5356 rings have a much higher pitch that's easy to hear. The 4043 sounds quite dull in comparison using this simple test. Some filler rod manufacturers now imprint the alloy along the length of the rod (Figure 9-69). This is great…if you can read text that small!

Figure 9–70 Using rings for eyebolts.

Figure 9–71 Using rings for loops.

Fabricating rings. Many times a fabricator is called on to make rings of different sizes for various purposes. They might be used for tie down loops or eyebolts (Figures 9-70 and 9-71) or welded to a plate to make a quick pad, or lifting-eye. A simple way to make these is to use heavy stainless spring-type lock washers (Figure 9-72). The twist is taken out of the washer and the joint welded to form a nice round ring of heavy gauge material (Figure 9-73). It's quick and beats winding heavy rod for one or two rings.

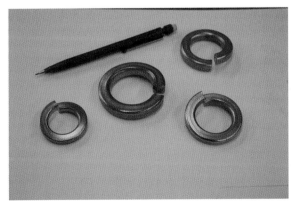

Figure 9–72 Working with washers.

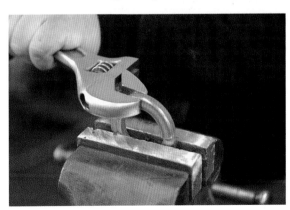

Figure 9–73 Forming a round ring.

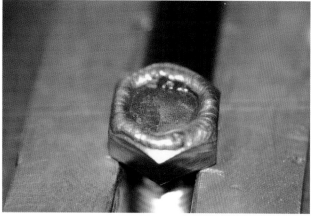

Figure 9–74 Creating composite materials.

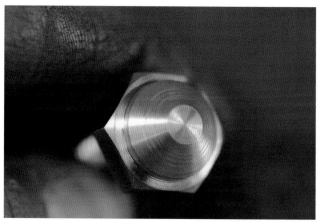

Figure 9–75 A specially-made centering screw.

Special materials. Using your brain and a little welding, you can create some unique composite materials for special applications. One I have used with success is to create a wear surface between two stainless steel parts by buttering one with silicon bronze weld buildup, sometimes called Everdur in the welding industry (Figure 9-74). This composite is then machined to create a sliding interface between the two parts. The example in Figure 9-75 was a special centering screw made from a 316 stainless bolt which was then built up with silicon bronze and machined to size.

Another useful trick involving copper is for heat sinks. If you clamp a copper plate behind a gap you want to fill, it gives you a backing that the weld metal lays on without sticking (Figure 9-76).

The copper backing bar absorbs a large amount of heat that would normally go directly into what you are welding, causing additional distortion. This is particularly helpful with thinner materials. The backside of this example is flush with the parent metal and the top surface has an almost flat contour that makes for minimal grinding and no weld prep (Figure 9-77).

Figure 9–76 Using a copper plate for backing.

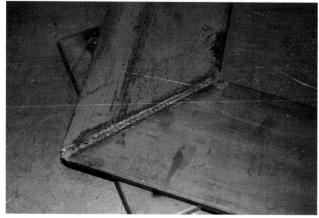

Figure 9–77 Minimizing grinding and weld prep.

Figure 9–78 Comparing angle frames.

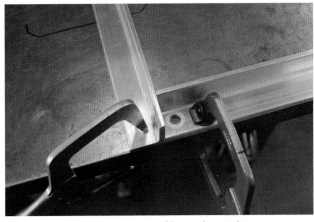

Figure 9–79 The advantages of butted straight cuts.

Minimize welding prep by reviewing your design. A few simple changes result in less weld preparation and easier fitup. The angle frames in Figures 9-78 and 9-79 are good examples. They are almost functionally equal. The miter (Figure 9-78) takes longer to cut and requires weld prep for a decent weld. The butted straight cuts (Figure 9-79) are pretty easy to prepare and produce a frame more quickly.

Figure 9–80 Structural beams.

Structural beams. Structural beams are a real chore to cut notch and fit up in certain ways (Figure 9-80). Here are a couple of examples of an easier way to fit up these types of materials.

With the simple addition of flat plates, straight cuts are used to join these beams (Figures 9-81 and 9-82). The plates are easy to cut and add strength to the connection by increasing the amount of weld in the joint compared to the notched fitup shown in Figure 9-80.

Figure 9–81 Using flat plates.

Figure 9–82 Joining the beams.

Sub-assemblies. Try to think in smaller sub-assemblies. They are easier to handle for welding and have a smaller distortion control problem. The leg sub assemblies seen in Figure 9-83 will be completed and straightened before they are added to the larger assembly. Each time two parts are joined by welding, there is some movement. If you don't correct for these small but cumulative effects, the main assembly will be difficult or impossible to correct. Small corrective action frequently is better than trying to make large corrections all at once.

Figure 9–83 Leg sub-assemblies.

Tube fitup. Deep butt weld tube fills on heavy wall material can be a real welding chore. To get a nice flush weld that grinds smooth takes a large amount of welding and filler material (Figure 9-84). An alternate that looks great and saves on welding and the subsequent grinding and straightening is to use a tube one size smaller and fillet weld (Figure 9-85).

Figure 9–84 Getting a flush weld.

Figure 9–85 Changing tube size and weld.

Another way to handle tube-to-corner fills is to inset the member away from the radius. (Figure 9-86). By insetting the tube away from the diverging formed corner, the gap becomes smaller. Moving the tube in leaves us with a much smaller, less distorting weld to deal with. My minimizing the weld size we are also directly controlling and minimizing the weld distortion. If the tube must be set flush with the outside surface, a small length of round material can be laid in the groove to help minimize the weld size if allowed.

Figure 9–86 Working with tube-to-corner fills.

Figure 9–87 Putting threaded bosses in a tube wall.

Figure 9–88 Using long bolts and studs.

Thread inserts. Sometimes a fabricator is required to put threaded bosses in a tube wall. Figure 9-87 shows a simple way to handle that without making any special fittings. The hole size is the dimension across the points of a standard or heavy nut. You can hold the nut for tacking using a long bolt. This beats drilling and tapping round bars in the lathe to make custom weld nuts. Be sure to use a long bolt or stud when you install these in a frame (Figure 9-88). With the long stud or bolt, it is easier to get the thread axis square with the world. You can make your own weld nuts obviously, but these are readily available and dirt cheap compared to the home-made models. How many threads do you need anyway?

If the hole is sized correctly, then installation is a snap. Remember: it's easier to sand a little off the corners of the nut than to produce a hole that fits the nut perfectly. For sealed threads, we use heavy nuts and weld a sheetmetal punch slug over the end of the nut (Figure 9-89).

Another use of this technique is to use your entire tubing frame as an air storage tank. I built a frame one time where all the tubes were interconnected for the purpose of an air storage tank. The frame had a lot of pneumatic equipment on it, so pipe thread bosses were installed wherever air was needed on the machine. It was quite a large frame and had a considerable air volume inside the tubing.

Figure 9–89 Sealed threads.

Figure 9–90 Changing an annoying weld...   *Figure 9–91* ...to an easy weld job.

Simplified weld joints. A simple change makes an annoying welding job (Figure 9-90) a piece of cake (Figure 9-91). Joint access is something that is difficult to predict for designers, but even harder to execute by welders.

Long wire and electrode stick outs are needed to reach into this annoying-to-fill joint (Figure 9-92). The tube wall takes the brunt of the weld heat and penetration; on thinner tubes, you risk burn through on the tube ID. It's not too bad with MIG and stick welding, but is particularly important with TIG welding thin wall tubing in stainless steel where you don't have the reach.

A small change to the welding joint can make anybody look like a professional welder. A simple rearrangement of the parts can save you time and make your welding look like show quality (Figure 9-93).

Figure 9–92 Using long wire and electrode stick outs.   *Figure 9–93* Adjusting the joint design.

Figure 9–94 Two formed channels.

Figure 9–95 Leave the bend open a few degrees.

Make your own custom tubing. The more professional looking method is to use two formed channels (Figure 9-94). When forming the channels, leave the bend a few degrees open to allow for the weld contraction (Figure 9-95). It's easier to flatten a protruding weld seam than to pull up a sunken one. This is sometimes the only way tubing can be obtained in special materials or odd sizes. The trick works well for all those small quantity, short lead time jobs.

Be sure to check squareness when you are tacking the channels together (Figure 9-96). With the welds off the corners, the custom tubing has that professional high production look just like the commercial material (Figure 9-97).

Figure 9–96 Check squareness.

Figure 9–97 Custom tubing.

You can also make tubing from two L-shaped formed parts, but the results never quite look like real tubing.

Form to eliminate or at least minimize welding. If you have a choice, form the longer runs instead of welding them, as shown in Figure 9-98. Remember the rule, "Form long, weld short."

Figure 9–98 Form long, weld short.

9.5 Brake Bumping

You can bump roll cylinders and other radii easily in the press brake. Thick-walled tubes and rings can be formed by this simple-to-learn technique. Many folks have seen large light poles out in public that are formed this way. They show the characteristic cogging or faceting that this method produces sometimes. Fundamentally the method is no different than rolling in a set of slip rolls. With slip rolls, the material is formed in a continuous fashion. Every tiny bit of material is bent by the rolls to form the smooth curvature. When we bump roll in the press brake, we abbreviate the process by only forming incrementally to get an approximation of the full curve. Depending on how many times you care to bump, the end results are indistinguishable from a rolled form. The advantages are that press brakes generally have a capacity for heavier or longer work than rolls and have no real lower limit on diameter. The disadvantages are seen when bumping large one-piece diameters because of the opening of the press brake. The one exception is a press brake equipped with a horn extension.

I start by laying out a series of parallel lines one inch apart (Figure 9-99). These general guide lines will be used to reference your progress. You will also need a sheet metal sweep or template to check your progress as you go (Figure 9-100). I always start out with three hits per inch with a light bend. You can calculate the approximate angle of each bend if you feel like it, but I usually just use a test piece to get the radius close. Keep in mind it's easy to add more curvature, but a real pain to take some out. Therefore, always come up a little short with three hits per inch.

Figure 9–99 Laying out parallel lines.

Figure 9–100 Checking progress.

To correct final curvature, you can bump in between your previous hits to add more curvature instead of increasing the amount of each individual bend. This produces more predictable results than increasing the bend angle by deeper punch depth.

This method of brake bumping is also used to pre-form the ends of plates and sheets that are going to be rolled in conventional rolls (Figure 9-101). With many power rolls, there will be a large flat spot at the ends of the plate if they are not pre-formed.

Figure 9–101 Brake bumping.

Figure 9–102 Forming perpendicular strips.

The strip on the left side of Figure 9-102 is used as a rough guide to make sure the strip is formed perpendicular to the edge. Typically rings are formed in two halves when they are a large diameter. Because of the way the material is supported in the press brake, we must add some on each end so the end can sit on the die for the forming process. This excess will be removed later when the part is formed to the correct radius. For large or heavy parts, you can suspend the part with a sling from the crane using a clamp or a choker hitch.

The flanged, one-piece cylinder in Figure 9-103 was formed with the press brake bumping method. The flange was formed on the flat pattern in the press brake and hung over the end of the die as the cylindrical portion was formed. You can see the faint lines from the bumping on the ID (Figure 9-104). This kind of cylinder can be formed in a set of rolls, but only if they have a deep relief groove in the top and bottom rolls, which most rolls don't have.

Figure 9–103 A flanged, one-piece cylinder.

Figure 9–104 Faint lines from bumping.

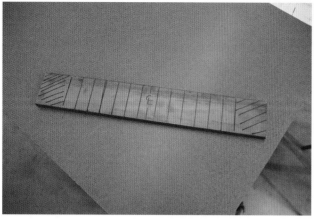

Figure 9–105 Leaving extra material on the cylinder blank.

Figure 9–106 Preparing the cylinder material.

Extra material is left on the ends of the cylinder blank in Figure 9-105 so the material can sit on the wide lower die opening properly for bumping (Figure 9-106). The forming sequence is to form in from each end first (Figure 9-107), and then form the center section last (Figure 9-108). It is much easier to position the part for the multiple strikes used in cylinder bumping if you use this sequence.

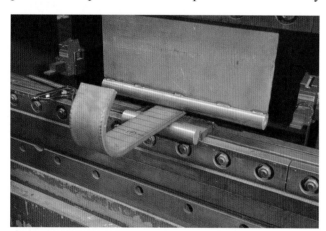

Figure 9–107 Form in from each end first.

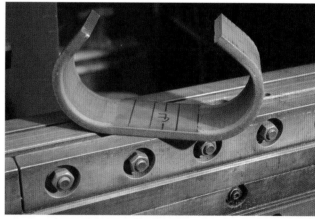

Figure 9–108 Slip the cylinder off the punch.

After forming, the cylinder is slipped off the special shop-made deep punch. I mark the trim points with a center punch before I form the blank. After it is partially formed, it's easy to see where the cut will be made. The extra material that was added for spanning the lower die is now trimmed off (Figures 9-109).

Figure 9–109 Extra material is trimmed.

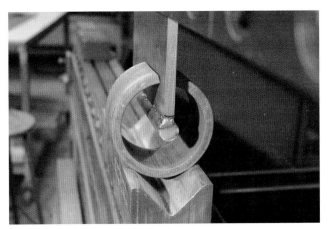

Figure 9–110 The cylinder is closed.

Figure 9–111 The tube is tacked together.

The cylinder is closed (Figure 9-110). After the last couple of hits are made to round up the tube, it is slipped off the upper punch and tacked together (Figure 9-111). Material is .50 thick × 4 wide steel. You can fabricate these custom-sized short tubes that are hard to even find commercially, let alone purchase in short lengths. Figure 9-112 shows a special sleeve made to fit an odd-size tube.

Extra-wide lower dies for bumping can be quickly made in the fabrication shop for a rush job (Figure 9-116). Figure 9-117 shows a custom lower die made for bump forming heavy plate.

Figure 9–112 A special sleeve for an odd-size tube.

Figure 9–113 Extra-wide lower dies.

Figure 9–114 A custom lower die.

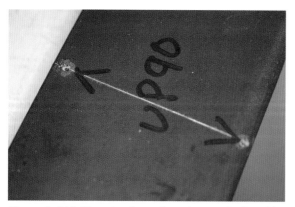

Figure 9–115 Use punch marks to locate the bend lines.

Figure 9–116 Center punch the ends of the bend lines.

Punch marks. When working with heavier material in the press brake, it's sometimes easier to use punch marks to locate the bend lines than program or set the back gage (Figure 9-115). Most back gages in plate shops are beaten to death from heavy plates banging into them. Center punch the ends of the bend lines (Figure 9-116) and just hide the punch mark under the upper punch (Figures 9-117 and 9-118) to locate the bend centerline. Marker smears and mill scale flakes off during forming, taking any scribe lines with it. Center punch marks will still be there.

Just barely hide the diameter of the center punch mark (Figure 9-118). This puts the center pretty darn close to the punch centerline.

Figure 9–117 Hide the punch mark.

Figure 9–118 Hide the diameter of the center punch mark.

Capping tube ends. Here are three methods of capping tube ends in order of speed and preference from worst to best (Figures 9-119, 9-120, and 9-121). The cap fitted for a corner weld has the advantage that it can be finished by grinding to make an invisible connection that looks like the stock tubing (Figure 9-119).

Figure 9–119 Capping tube ends.

Figure 9–120 Fitting the end cap outside.

Figure 9–121 The fastest capping method.

Sometimes it is easier for fitup if the end cap is designed to fit on the outside (Figure 9-120). Fitting the end cap to the inside requires more time and higher precision in each of the two parts, not to mention actually positioning it for welding. The corner-to-corner method requires weld finishing if a blended look is desired. The method in Figure 9-121 gives a larger footprint, easy welding, and low precision of the mating parts.

Frames. It may be more efficient, depending on the available equipment, to make a picture frame or large window cutout by connecting flat strips for the outline (Figures 9-122 and 9-123).

Figure 9–122 Making a frame or window cutout.

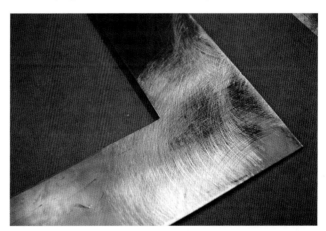

Figure 9–123 Connecting flat strips.

Cutting a large center out of a flat sheet is more wasteful of material than shearing strips and can have stresses left in depending how the cutting was done. It's hard to beat the clean, neat look of sheared edges on the inside with almost any manual cutting process.

If you make the frame a little large, you can actually shear it to final accurate size after welding and grinding (Figure 9-124). With good sound welding, forming this frame into a door or cover is a simple matter.

Figure 9–124 Shearing a frame if needed.

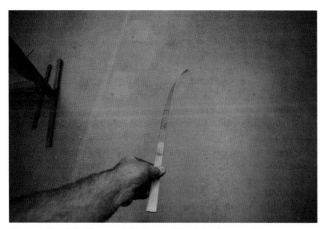

Figure 9–125 Sheared Strips get twisted sometimes.

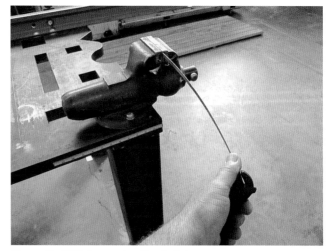

Figure 9–126 Removing the twist.

Twists. Many times when you shear a narrow strip for a project, the shear puts a long sweeping twist in the material (Figure 9-125). The best way to remove the twist is to hold the strip by the extreme end with minimum material clamped in a vise (Figure 9-126). By twisting it as a whole with a wrench from the extreme opposite end, you get a perfect straight strip. The shear twists the entire strip when it's cut, so the correction must be the precise opposite of the shearing process. If you clamp short sections and work your way down the strip, it will take longer and the results will be pretty dodgy.

A similar trick works for straightening wire. Clamp one end of a long wire you want to straighten in the vise. Clamp the opposite end in a drill motor or loop it around a small eyebolt held in a drill (Figure 9-127). Pull a little tension on the wire and spin the drill (Figure 9-128). What happens is the wire yields as it is spun and takes on the new shape, which, if you did it right, is nice and straight. A common use would be straightening small diameter MIG wire for use as TIG filler rod.

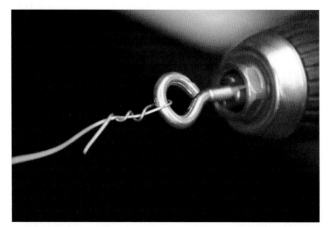

Figure 9–127 Straightening wire.

Figure 9–128 Spinning the drill.

Figure 9–129 A squared and braced frame.

Figure 9–130 Clamping braces.

Squareness. After you have squared and braced a simple frame (Figure 9-129), be sure to use the proper weld sequence to maintain your squareness. Always make the least distorting welds first and check for movement *as you go, not after* you have welded it up. Braces can be tack welded or clamped (Figure 9-130).

Figure 9–131 Typical tack welded joint.

Figure 9–132 Ending on a tack weld to avoid movement.

Always end on a tack weld instead of starting on one (Figures 9-131, 9-132, 9-133). If you start on a tack weld, it softens it and allows movement before the joint has developed some strength of its own from your weld.

Figure 9–133 Not heating the tack gives the joint time to strengthen.

Figure 9–134 Welding together two pieces of flat bar.

Figure 9–135 The gap between two pieces.

Fillet welding. In general, leave all fillet welding until last. Fillets distort terribly, but most people like to do them first because they are easy and look good.

Never weld anything in an unrestrained condition unless you want it to move. The lowly clamp has the ability to reduce your welding distortion by a huge amount.

An easy way to think about it is if you are welding two pieces of flat bar together end-to-end with a gap between them, as shown in Figures 9-134 and 9-135. In the first picture, the two bars are clamped lightly with vise grips, so I could measure the overall length without the bars shifting. In the second picture, I have clamped a second set of bars with a pair of heavy duty, professional welding clamps. Machine settings are identical for both welds.

If you can imagine the weld metal as a tiny turnbuckle that pulls the ends of the flat bars closer together as the weld metal cools, you are halfway there to understanding what takes place (Figure 9-136). Now visualize adding a clamp on each bar that will not allow the bars to move closer together. The result is that all the weld shrinkage takes place within the weld metal instead of using its shrinking energy to pull the bars closer together. From this simple test, you can see the longitudinal shrinkage was reduced by half just by changing the type of clamps that were used (Figure 9-137). If you were really paying attention, you would have noted that even under heavy clamping it still moved. Any good welder has learned how to deal with and predict weld distortion. If you remember the simple credo, "Everything moves, always," you're halfway there.

Figure 9–136 Pulling the ends closer.

Figure 9–137 Reducing longitudinal shrinkage.

Figure 9–138 Checking parallel dimensions.

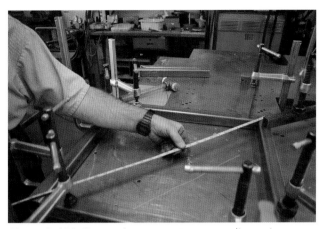

Figure 9–139 Comparing corner-to-corner dimensions.

Squaring. Check parallel dimensions first when squaring anything (Figure 9-138). You cannot be square if you're missing parallelism first. By comparing the corner-to-corner dimensions, you are using a sensitive geometric proof to verify a square (Figure 9-139). A combination square or carpenter's square is a more difficult tool to use than the cross corner method when moving at speed for squaring jobs.

Accurate surfaces. If you have a fabrication that needs several accurate surfaces, it's easier to weld a pad for the precision component to sit on (Figure 9-140). This saves having to grind or sur-

Figure 9–140 Welding a pad.

face the entire face and the inherent problems with that method. Don't make the mistake and skimp on the machining or grinding allowance of your component pad.

If you need a continuous precision surface on an otherwise lightweight fabrication, attach flat bars to provide material to precision surface for flatness of other geometry (Figures 9-141 and 9-142). Generally, tube walls should not be surfaced directly to create a precision surface because of excessive wall thinning. These flat bars will be surface ground after the frame has been stress relieved to provide a super-accurate mounting surface for some fussy components (Figure 9-142).

Figure 9–141 Attaching flat bars.

Figure 9–142 Providing an accurate mounting surface.

Figure 9–143 Angle iron V-block.

Figure 9–144 These V-blocks are easy to make.

V-blocks. Make yourself a few sizes of angle iron V-blocks (Figures 9-143 and 9-144). These are easy to make and handy around the welding shop when working on piping or handrail projects Make a few sizes in pairs for all those round fabrication projects in the shop.

For long V-blocks, you can use old sections of press brake dies. The extra length makes a big difference when you have to connect two longer rounds together (Figure 9-145). Don't use nice new precision dies unless you want to create an enemy in the forming department. Brake dies are precision tooling and should be treated as such.

Figure 9–145 Connecting longer rounds.

Figure 9–146 An alternate way for prepping the ends of rounds.

Connecting rounds. Figure 9-146 shows an alternate way to prep the ends of rounds to weld them together. This is the preferred method to avoid lots of stops and starts in your welding. By prepping the round with a standard double bevel, you can do most of the fill without rotating the part every few degrees. For welding processes that produce slag, fewer starts and stops are important. In big reinforcing rod joints, the weld preps are done this way.

Delicate work. Clamp your ground clamp directly to delicate work (Figure 9-147). Some materials are very sensitive to arc erosion through the ground—aluminum and magnesium are two examples. These marks look really bad and are hard to remove because there is usually metal transfer from the table to the part (Figure 9-148).

Figure 9–147 Connecting ground directly.

Figure 9–148 Some materials are very ground sensitive.

Auger layout. Figure 9-149 shows the start of an auger layout. The long centerline is marked on the tube by laying the tube in a length of angle and scribing or marking along the edge. The pitch is marked off along this centerline to guide the tacking and fitup of the flighting sections (Figure 9-150). On larger screws and augers, a come-along or crane can be used to help stretch the flighting along the center tube.

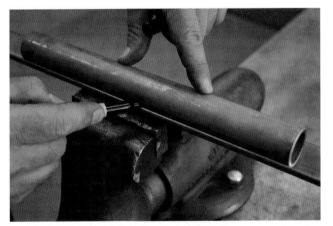

Figure 9–149 The start of an auger layout.

Figure 9–150 Guiding the tacking and fitup.

Pre-heating. If you have some heavy material to weld, you can reduce your amperage requirements or increase the maximum thickness you can weld with some judicious pre-heating (Figure 9-151). I like to preheat anything over 3/8 thick when we are running smaller wire diameters in the MIG welder. It helps with welding speed, fusion, and re-starts. My temperature gauge is when I see a little smoke from the surface scale or a temperature that will cause leather to smoke a little when touched to the metal.

Figure 9–151 Judicious pre-heating.

Figure 9–152 Stainless filler rod.

Figure 9–153 Precision alignment and fitup.

Filler rod. You can use stainless filler rod to tack up a steel job, if the job will allow it (Figure 9-152). The stainless tack is much more ductile than a steel filler tack. The parts can be easily bopped around with a hammer to allow precision alignment and fitup (Figure 9-153). The advantage is that stainless tacks are the more ductile material and don't crack like steel filler when you make fitup adjustments.

Porous casting. I'm sure every welder has fought a porous casting at one point or another (Figure 9-154). Sometimes nothing you do will clean out the embedded contamination and pinholes to allow a decent weld. One trick I have used with success is to use a drill to actually drill into the offending pockets, kind of like a dentist drilling out a cavity in your tooth (Figure 9-155). The drill removes the buried pockets of crud and gives you a fresh shot at making a decent weld. Another trick along these lines, if you have the facilities, is to cook the casting at a high temperature to leech out the dirt and crud. The temperature will vary depending on the material you are working. As a rule of thumb, only cook the part at a temperature at which you would do a preheat, or maximum inter-pass temperature.

In this chapter, we looked at many different tricks and techniques used in the welding shop to handle those tricky jobs and work more efficiently. In the next chapter, the discussion will move into methods used to control welding work to a much higher degree.

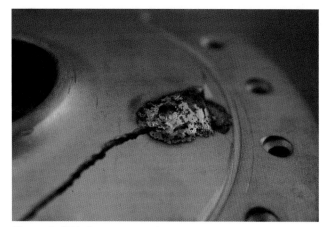

Figure 9–154 A porous casting.

Figure 9–155 Thinking like a dentist!

The Lost Art of Flame Straightening

10

10.1 Limitations
10.2 How Flame Straightening Works
10.3 Heat Input
10.4 Mapping
10.5 Applying the Correction
10.6 Straightening Shafts and Tubes
10.7 Special Applications of Heat Shrinking
10.8 Correcting Weldments

Flame straightening is one of those old-timer skills everybody has heard of but few people have had the opportunity to watch someone do. The ability to straighten bent plates and shafts is almost magical in its simplicity (Figure 10-1). It's definitely a valuable skill to learn if you do any significant welding, unless you relish the though of beating a stubborn shaft or plate into submission with a mallet. This easy-to-learn technique using simple shop equipment has been used to straighten everything from the hulls of ships to samurai swords.

Welders and metal fabricators are continually confronted with the negative aspects of joining metals by welding, some welders more than others. Whenever metal is heated it expands and not always in a favorable way. Most often it has a negative effect on the end product. We have all seen the results manifested as warpage, distortion, and stress in our own work.

Figure 10–1 Straightening bent plates and shafts.

259

If you have spent any time in a weld shop, you have seen some of these kinds of distortion like welds breaking, flame cut strips and shapes twisting and bowing, oil canned plates, and covers. The list is as long as there are things that are heated during manufacture.

Every welded structure will try to seek the position of least stored stress. A simple example that best illustrates this is when you crumple up a sheet of paper tightly and then drop it on the desk. The paper expands slightly and finally settles in the position of least internal stress.

As we weld on a metal structure, this is happening continuously as we work. The weldment is undergoing a constant re-arrangement to find this state of least stress. The welding sequence, clamping, and fixturing can help, but can never eliminate the effects of expanding hot metal. When we stress relieve a weldment or a part, we are allowing whatever remaining locked-in stress to dissipate by heating or, in some cases, vibration.

Flame straightening uses the same principles that cause the warpage and distortion to make positive corrections or, in some cases, enhance the effects. The best way to imagine how this process works is something like reverse welding.

Figure 10-2 shows an example of using the effects of controlled distortion to induce a large camber. These kinds of results can be accomplished with the humble cutting torch and the knowledge of where to apply the tool. In this example, if I had a set of rolls large enough I would have run the tubes through. These kinds of results can be achieved in a one-light-bulb garage or standing on scaffolding five hundred feet off the water.

Unlike pressure methods like hydraulic presses and screw jacks, flame straightening is the laser-guided smart bomb that attacks the exact offending

Figure 10–2 Inducing a large camber.

areas. Pressure methods distribute their gross application of forces throughout the structure and sometimes cause more harm than good. Hydraulic rams further complicate the correcting measures because they lack a sense of feel connected with the yield point of a particular material. Unless you have a good pressure gauge to monitor your force application, the yield point can be a dangerous, easily-exceeded target.

Pressure methods require the parts to be supported so the force can be applied. It's kind of like pulling yourself up by the bootstraps. I have seen good work ruined by the sloppy application of force methods. I have also seen a guy knock out his two front teeth using a hydraulic ram to straighten a frame, but that's a story all on its own. In many cases, the materials physically cannot bear the loads needed to restore them.

Force methods typically require increased operator handling and labor. With flame straightening, plates can be straightened installed as the hull of a ship or where they lay in the shop. Shafts and tubes can be corrected while they're still in the lathe or connected to machinery.

The portability of this method alone is worth the time spent to learn it. Flame straightening can be used where pressure methods would be impossible or extremely awkward. This is not to say that pressure methods don't work. As any experienced metal fabricator can tell you, sometimes you need all the tricks in your bag to get the job done. All I am doing is giving you something else to put in your bag.

So what is flame straightening? I like to define it as *the specific application of controlled cycles of heating and cooling that are used to correct, enhance, or minimize distortion in metals.*

Figure 10–3 Bowing the long tube.

Flame straightening causes a contraction or localized shrinking in metals that are used by the applicator (you) to affect changes in geometry. The technique is applied with a common oxy-acetylene torch in a specific way that induces the maximum amount of contraction. Before you run out to your shop and fire up your torch, you should read the rest of the chapter. There are a few important points you may want to know about.

Flame straightening works best when the distortion is caused by welding or is heat related. It can be used on any materials that have no restrictions on the application of heat from an oxy-acetylene torch. It works best with materials that do not have high thermal conductivity. Steel and stainless steel are the primary candidates.

Aluminum and copper are less than enthusiastic about this technique. You can see in Figures 10-3 and 10-4 the two fillet welds that have bowed the

long tube. This would be a typical weld distortion problem encountered in the shop. Some of this bowing could be eliminated by good fixturing and clamping, but for the example I wanted a maximum distortion.

Figure 10–4 A typical weld distortion.

10.1 Limitations

Flame straightening shouldn't be used on materials that would be damaged by the application of heat. Wood is not a good candidate. A few examples of materials you should avoid are metals that react with oxygen when heated (titanium) and heat-treated parts that have been tempered to temperatures less than 1300°F. Certain kinds of alloy steels, tool steels, and some steel forgings can be damaged by the kind of rapid heating used in flame straightening. In other words, be sure you know what you are working with before you screw something up. Don't say I didn't warn you! This is where welding engineers earn their keep.

10.2 How Flame Straightening Works

Metals, like all materials, expand when heated. In the case of steel, it expands .000006" per inch, per degree of temperature change. By using this technique of very localized, intense heating, we can cause what we are looking for—a small cross section of material to expand considerably and form an actual material upset, and then the subsequent mechanical contraction of the material as it cools.

We are trying to induce a large temperature difference between the area we are heating and the surrounding cooler metal. The contraction or shrinking effect we desire is created by the temperature difference or gradient. We are, in effect, creating the same conditions we see when welding: a very hot, intense, local heating surrounded by cooler base material.

As this local area is heated, it is trying to expand rapidly. It encounters resistance from the cooler surrounding material, which creates pressure. This pressure flows in the path of least resistance, which in most cases is upward along the area softened by heating. This upward swelling forms the upset. And as it cools, it contracts and acts like a tiny heavy duty turnbuckle pulling the surrounding material with it. This upset can be seen on thinner materials as a slight bulge or slight swelling. In Figure 10-5, I have intentionally highlighted the slight bulges left from the contraction by dusting them with a sander. Notice that the corrective action was done directly opposite the fillet welds on the tee which caused the distortion.

Figure 10–5 Bulges left from contraction.

10.3 Heat Input

One of the basic secrets to using this technique is the heat input. The trick is to heat the material as quickly as possible in as small an area as possible (Figure 10-6). If we used a normal rosebud-type heating tip, we cannot confine the heat to a small weld-sized zone. The best tips to use for most normal flame straightening are cutting tips. Only the preheat flame is used to form the upset.

These tips are designed to heat a small area quickly in preparation for flame cutting so they work well for the rapid heat input into a small area we are looking for. Rosebuds are for burning weeds in back of the shop or large area heating, not tight localized spot heating.

Figure 10–6 The heat input is key.

For this example, I had some mechanical assistance from an aluminum strong-back and a couple of clamps (Figures 10-7 and 10-8). This enhanced the effects of the heat shrinks so the work was completed faster.

Figure 10–7 Help from a strong-back and clamps.

Figure 10–8 Enhancing the effects of the heat shrinks.

Figure 10-9 shows the sample after correction. The material is $2 \times 2 \times 1/8$ wall steel tubing. I am pointing to the lack of a gap between the straightedge and the tube.

Always use the correct fuel and oxygen pressure and flow for the size tip you are using. There is some confusion as to how far to open the fuel valve when setting your flame. Open the fuel valve and light the torch in the normal manner with the correct regulator pressure settings for the tip size you are using.

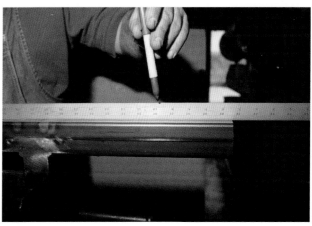

Figure 10–9 The sample after correction.

Figure 10–10 Opening the fuel valve.

Figure 10–11 Adjusting the fuel valve.

After you have it lit, open the fuel valve until there is a separation between the base of the pure fuel flame and the tip (Figure 10-10). Slowly close the fuel valve just until the base of the flame re-connects with the torch tip (Figure 10-11). This is the correct amount of fuel for the tip. More than this is too much fuel flow and the flame will be hard and forceful. Less than this and you are not heating as efficiently as you could be.

We always apply the corrective heat to the high spot we are trying to remove. If it is weld distortion, we work on the opposite side from the offending weld. The only exception to this is if we were trying to induce curvature or deliberate distortion such as a camber.

10.4 Mapping

The first step in any correction job is mapping the high points with a straightedge (Figure 10-12). For the demonstration, we will map and correct the distortion in a typical mill supplied plate.

You will need a few tools to help you map the distortion. Several straightedges of different lengths and markers that won't wash off with water are used to plot the typical wandering distortion found in plates. I like the faithful old Sharpie. It leaves a hazy reminder of where you have applied shrinks even after the intense heating.

The first step in plotting the distortion is to do a quick check with the straightedge along both axes of the plate to find the worst side—the side with the largest number and most severe high spots.

For correcting plates, I like to have the plate on horses or blocks on the table. Support the plate on three points so it doesn't rock. The best is outside on the forks of a forklift with a water hose nearby.

There is a sweet spot where you don't have to bend too far to sight the gap between the straight edge and the plate is not too high to reach across. The older you get, the narrower this range is.

To locate the high spots, we will use our longest straightedge first. Hopefully it spans the entire width of the plate. If you don't have a straightedge long enough, a decent piece of flatbar will also work. Before you begin, decide how flat you are going to make the plate. It's pretty easy to change your standards as you work and end up putting more work in than necessary.

Figure 10–12 Mapping the high points with a straightedge.

Figure 10–13 Plotting in the direction of worst distortion.

Figure 10–14 Comparing the straightedge to the plate.

Start your plotting in the direction of worst distortion. Move your straightedge perpendicular to the edge (Figure 10-13). If the plate is round or odd shaped, pick a direction and stick with it for the entire straightening job. As you compare the straightedge to the plate, place your eye at the same level as the gap between the plate and the straightedge (Figure 10-14). This makes it easier to see the gap.

Move the straightedge along the plate. As you find the high spots, mark them with your marker. Mark in approximately two-inch increments as you move along the plate. When you find a high spot, rock the straightedge back and forth to locate the apex. The rocking see-saw motion will show you exactly where the apex is. Mark these apexes as you move along the plate. If there are two high spots or the high spot is broad, mark the points where the change is greatest.

It is very common to have distortion that requires you to flip the plate from side to side. Do as much correction on one side as possible before you flip the plate. It's much too awkward to try to apply your correction in the overhead position. Unless you have something to prove about your abilities as a contortionist, I would just flip the plate. If the plate weighs 25 tons, then maybe it makes more sense to do it from the underside.

I got to pick this example plate and, in case you hadn't noticed, it's pretty small and easy to flip for us old guys.

As you make your marks along the plate, use line weight to remind you of the approximate amount of distortion. A light dash indicates a minor variation and a heavy dash indicates a large distortion requiring a full correction. Many times when mapping the distortion, you will find that it fades out and ends abruptly. Mark the end point to remind you to stop your straightening at that point (Figure 10-15). Typically there will be several wandering dashed lines, some ending and some running the full length or width of the plate.

So now we have marked out the variations in the plate we want to correct. We are now ready to apply our correction.

Figure 10–15 Marking the end point.

Figure 10–16 A full heat shrink on steel.

Figure 10–17 Straightening a stainless steel plate.

10.5 Applying the Correction

The flame straightening correction is called a "heat shrink." It is applied with a common oxy-acetylene torch with a standard cutting type tip. Tip selection is determined by the material type and thickness. Use a tip the same size you would for cutting when shrinking stainless steel and one size larger than you would for normal cutting for steel. Oxygen Acetylene is the preferred fuel combination. Acetylene oxygen has a higher Btu/min input than other common cutting fuel combinations. Remember we are trying to heat a narrow zone very quickly.

When we make our corrections, the heat shrink is applied in the same manner we would use for welding. Uniform travel speed and standoff distance are important for getting consistent results. The inner pre-heat cones of the flame should be held off the surface to prevent scarring.

A full heat shrink on steel, as shown in Figure 10-16, will have the entire width of the shrink heated to a low red or cherry red heat approx 1000–1300°F. In Figure 10-17, you can see I am straightening a stainless steel plate with a very light shrink.

Stainless steel is heated only until a deep brown is seen on the surface, less than 900°F (Figure 10-18).

Because stainless steel is a poor heat conductor and can be damaged by overheating, we use a lower temperature when straightening stainless. Stainless steel expands about 1 1/2 times as much as steel, so less aggressive shrinking is necessary to get the same results. It's best to go easy and add more correction than to have to correct for over-zealous application of shrinking. In Figure 10-18, you can see I am straightening stainless with barely a heat mark visible.

An important point is that once you have heated along a line with a full shrink, no useful gain will happen by going over the exact same line, unless less than a full shrink was used. Normally if more correction is required, it is done along side of the original shrink line.

Figure 10–18 Barely a heat mark visible.

Here's a rough rule of thumb: If the plotted distortion 12 inches from the high spot is 1/16 or greater, the width of the heat shrink is equal to the material thickness. For stainless, reduce this by half for best results.

The heat shrink begins at an edge or in a single spot. For a full shrink on steel, the spot is heated until a dull red can be seen. The torch can now be moved along the length of the area to be straightened. It is important to maintain the temperature consistently as you move across the work. Remember, it's like reverse welding. Travel speed and standoff distance are as important as they are in welding. Weaving is recommended for a wider shrink area on thicker plates.

There are limits to how much correction you can get from heat shrinking. Typically you would see these near edges where plates are easily damaged in handling. It may not be possible to correct these types of damage without the use of aids like clamps or other mechanical aides. The effects of corrective heat shrinking will always be enhanced by the application of mechanical assistance. Weights and clamps will greatly magnify your heat shrinking results.

Figure 10–19 Cooling the shrink.

Figure 10–20 Using a hose.

Cooling the Shrink. How we cool the heat shrink is almost as important as how we heat the work in the first place. Eventually the heat we input into the part will reach some equilibrium point if we leave it alone long enough. The effects of the heat shrink can be magnified by effective cooling of the heated metal. The cooling effectively freezes the upset from the heat shrink and gives us more correction per shrink than if no cooling is used.

For small jobs, a spray bottle of water or even a wet rag can be used (Figure 10-19). For larger jobs, a stream from a hose works the best (Figure 10-20). Remember that an accurate assessment of the correction cannot be done until the part is cool enough to put your bare hand on (Figure 10-21). Additional shrinking should not be done in the same area until the part is cool enough to touch. Compressed air can be used to blow off excess water so you can see the gap between the straight-edge and the part.

Figure 10–21 When is the part cool?

Cooling should start on the hottest part of the shrink first. This helps freeze the shrinks in their new contracted positions. Thin tubing and sheet should be cooled immediately after the heat is removed.

It is okay to apply several heat shrinks in one heating and cooling cycle, but they must be separated by cool metal to be effective. Heat shrinks should never be applied side by side without cooling unless two torches are used simultaneously for a wide heavy-duty shrink.

Figure 10-22 shows some of the other flame straightening plotting symbols I use. Most of

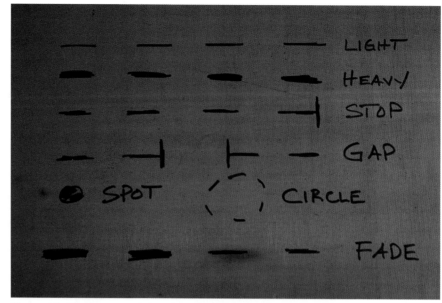

Figure 10–22 Flame straightening plotting symbols.

these are pretty easy to understand. The main thing to remember is that these symbols represent what we see when the plate or part is examined. If you trust your eyes and measuring tools, the symbols are only a reminder of what needs to happen for a particular section.

A light dashed line represents a minimal application for a minor correction. A heavy-handed mark means a full power heat shrink to me. A stop shrink would mean the damage or distortion ended at that point. The same would apply to a gap shrink.

10.6 Straightening Shafts and Tubes

Shafts and tubes are straightened using the same techniques as we use for plates. The easiest way to straighten a shaft is in V-blocks or, better yet, mounted between centers in a lathe with piped coolant available. How bent the part is determines what type of heat shrink we apply. With shafts or other cylindrical parts, we use either an axial or radial shrink. Excellent results can be achieved with this method. If you want to fuss around, you can easily get below .005 run out.

The axial shrink is the gentler of the two types of heat shrinks used on shafts and tubes (Figure 10-23). The radial shrink is for heavy duty distortion and should only be used if the axial type shrink does not produce the desired results. In the axial shrink, the heat is applied in the same direction as the shaft. Normally these are very short shrinks because the apex or high points on shafts are very localized.

A dial indicator can be used more easily than straightedges to accurately locate the high points on a shaft (Figure 10-24). These points are plotted and marked in the same way we would mark a flat plate. Because we mark the high spots, all of your reading will be plus readings. I use the center of the plus as my target when I apply the shrink.

Figure 10–23 The axial shrink.

Figure 10–24 Locating high points on a shaft.

The Aftermath of an Axial Shrink on a Solid Shaft. The high spot in this example was .008 inches (Figure 10-25). When we apply the axial heat shrink to a shaft, we start a little before the high spot and continued the same amount beyond the apex. The maximum length for any axial heat shrink should be below two inches long. If more shrinking is needed, then a radial shrink should be used. The shaft in this example is 1 1/2 inch cold rolled steel.

Figure 10–25 Finding the high spot.

Figure 10–26 Controlling a radial shrink.

For a radial shrink, the amount of shrinking is controlled by the radial distance that the shrink is carried around the circumference from the apex or high point of the distortion (Figure 10-26). For starting shrinks, about 20–30 degrees of radial arc would be shrunk. If the desired results are not achieved, more and more arc length is added up to a maximum of about 120 degrees or arc (Figure 10-27). If this is still not enough, additional shrinks can be added along side, or mechanical aides can be used to work a stubborn bend.

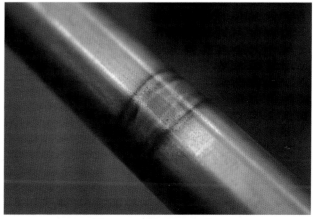

Figure 10–27 Working a stubborn bend.

Figure 10–28 Connecting lines to form a Vee.

Figure 10–29 Heating and shrinking both sides of the Vee.

10.7 Special Applications of Heat Shrinking

There are a few special cases worth mentioning here. The first is the high aspect ratio shrink. This is used for sections that have a greater depth than width, like rectangular or beam sections. Section depth limits how much correction we can get from normal shrinking of the apex of the distortion. For these deep sections, we need to involve more of the material. Remember the rule of thumb that the width of the shrink is roughly equal to the thickness or, in this case, the depth of the material. For these types of parts, we use a high aspect ratio or Vee shrink.

You can draw two lines separated by a distance approximately equal to the depth of the section, and then connect these lines together to form a Vee (Figure 10-28). This is the material we will shrink to correct this type of section. The Vee points away from the high spot of the correction. Typically both sides of the Vee are heated and shrunk (Figure 10-29).

When we heat this type of shrink, we start with the widest part and weave our way down to the point of the triangle (Figure 10-30). Mechanical aides such as clamps and weights can assist greatly in these cases. Cool the hottest part first. In Figure 10-31, I am using a spray bottle filled with water and a little soluble oil, which is the same coolant we use in the lathe.

Flame straightening in conjunction with mechanical aides allows straightening that would otherwise be impossible with flame straightening alone. Typically the assists come from clamps, jacks weights, and even gravity.

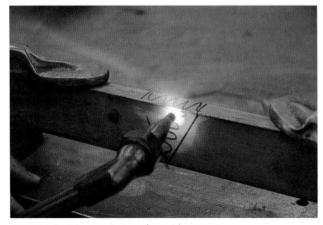

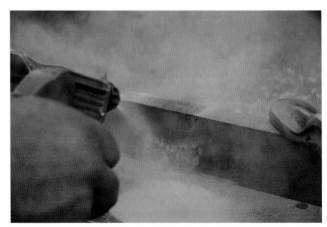

Figure 10–30 Starting at the widest part.

Figure 10–31 Using water and soluble oil to cool.

Figure 10–32 The challenge of weldments.

Figure 10–33 The bow caused from welding.

10.8 Correcting Weldments

Weldments present some of the most demanding and complicated uses of flame straightening (Figure 10-32). Due to the interaction of the different parts and pieces, analyzing where and how to make the corrections can be tricky. Just remember the distortions are most likely caused from the welding that goes into an assembly. In Figure 10-33 you can see the bow in the tube caused from the welding on the opposite end. Target your corrections to counteract these effects. It is also helpful to make smaller corrections to sub-assemblies before they are incorporated into a larger structure where the distortions can become impossible to correct.

In this example, heat shrinks were applied to the underside of the bridge beams to produce a camber which increases the load carrying capacity (Figures 10-34 and 10-35). I would be willing to bet this was done with these beams sitting on sawhorses using gravity for assistance.

I have outlined the basic techniques of flame straightening. The effective use of this procedure can only be learned by trying it and carefully observing the results. The only real requirement is the willingness to try it. Flame straightening is not the answer to every distortion problem that faces a metal fabricator. It is only another tool in the tool bag of a competent metalworker. The importance of good design and the use of proper manufacturing techniques are at least as important as knowing how to correct for and repair the defects that you encountered in your welding shop.

Figure 10–34 Producing a gentle camber.

Figure 10–35 Visible heat shrink marks.

Sheet Metal Shop

11.1 Layout Work
11.2 Blank Length Calculations
11.3 Patterns
11.4 The "Yank" Method
11.5 Forming and Layout of Cones
11.6 Tanks and Baffles

One of the old guys I learned a great deal from told me one time that a sheet metal worker's work was much more difficult than a machinist's work. I asked him to explain why he thought that was the case. (I had worked on both sides; I figured I could offer an answer that would help him appreciate the difficulties faced in machining.) His reasoning was a machinist began with a block of material and slowly whittled it until it was the desired shape and size whereas the sheet metal worker first had to decide the starting blank size and predict the final size after working.

I gently pointed out that I thought this was a gross over-simplification by an old codger who had never turned the crank on any machine more accurate than a drill press. After the lights stopped spinning and the ringing stopped in my ear, I thought about his point and could see where he could get this idea.

Sheet metal work is a precision trade without a doubt. If you don't believe me, try welding a badly fit-up job in 24 gauge stainless steel and see how far you get without tight accuracy and precision fitup. Typically, sheet metal work is not as closely controlled as machine work, even though some designers and engineers might think it is. Production type work can be tuned and adjusted to very close limits, but only at the expense of many parts. One sheet metal forming that is off can be tricky to do accurately on the first cut.

There are four basic types of sheet metal work: HVAC, Industrial, Electronic, and Aircraft and Automotive. They all overlap and the techniques learned in one class of work are easily applied to any of the others.

The first type, HVAC, involves Heating, Ventilation, and Air Conditioning. It is characterized by special seams and lock joints with transitions, little welding, and lots of duct tape and self-tapping screws.

Industrial sheet metal generally covers chemical and food processing equipment, conveyors, tanks, and vessels made from sheet metal. This work is in many cases liquid or pressure tight, sometimes in special corrosion-resistant materials with lots of welding, plumbing, and polishing.

Some electronic sheet metal is formed into the chassis and brackets used in computers and other enclosed electronic equipment. Electronic sheet metal has lots of punched or laser cut holes and cutouts, with inserted hardware, high accuracy, and minimal welding. Most forming is done in press brakes, sometimes using special tooling

Aircraft and automotive work covers sheet metal formed into compound shapes or warped three–dimensional surfaces. Much of this compound work involves more art than science at the individual worker level. It is easily the most complex work, where a solid understanding of material behavior and welding skill is required. It is also characterized by large amounts of hand work.

Obviously this is somewhat of a generalization on my part with many subsets of each basic type in actual industry, but it can be said that each basic types has its own set of typical problems and challenges. Keeping to the trend I think I have started in this book, I will focus on tips and tricks that cross over between the different basic types and have use to anyone involved in sheet metal work in the jobbing shop setting.

I define sheet metal as sheet material that is named by a gauge number of less than .188″ thick measured decimally.

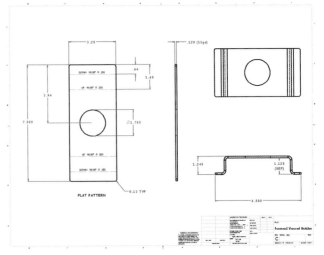

Figure 11–1 This drawing indicates the thickness.

For most sheet metal applications, it is less confusing for the designer and the shop personnel if decimal material thickness with tolerances and the intended thickness gauge are called out on engineering drawings. (Figure 11-1) I would classify material heavier than .188 as plate work, even though the same calculations and techniques are used.

More than any other trade, all sheet metal work starts with layout and planning—we must anticipate and predict how the material behaves before we touch tool to metal. Patterns and templates are commonplace. With the increasing use of computer pattern development and computer-controlled, high speed laser and water jet cutting, the need for expert pattern development is changing for some shops. During my early years, before computers, I found pattern development to be challenging and fun.

Classical pattern drafting is outside the scope of this book and is covered in excellent detail in many readily available books. I have listed several in the recommended reading section in the back of this book. So, what kind layout are we talking about? To me layout is the ability to plan and predict how the material will behave before you do it for accurate results.

11.1 Layout Work

Skip the bluing unless you really enjoy the smell. Dykem is for the days gone by before good carbide scribers. Press a little harder with your scribe. Save the drying time and the inevitable cleanup. In a pinch, you can use a little Sharpie marker to provide a better contrast on a light-colored surface. A Sharpie marker is used in Figure 11-2 to bring out detail on a difficult surface. Instead of coating the entire surface with bluing, with its inevitable cleanup, just cover the areas of detail (Figure 11-3).

There are several tools that can be used to mark a line parallel to another edge (Figure 11-4). For small dimensions less than an inch, a modified pair of dividers with a guiding edge and a scribing edge coupled with a screw adjustment is fast and accurate (Figure 11-5). The friction pivot type is quite lame and always seems to move just when you don't want it to move.

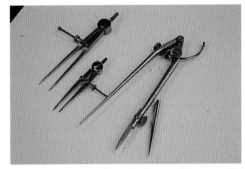
Figure 11–4 Handy tools for marking parallel lines.

Figure 11–5 Measuring less than an inch.

Make a quick scratch template from sheet metal for repetitive layouts (Figures 11-6 and 11-7). These two are 24 ga galvanized sheet metal. You can mark bend lines and hole centers for repetitive layouts.

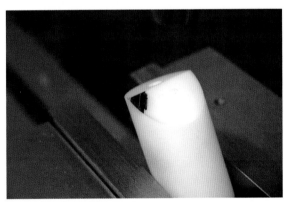
Figure 11–2 Bringing out detail.

Figure 11–6 A quick scratch template.

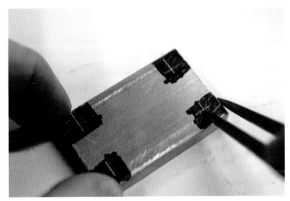
Figure 11–3 Covering just the areas of detail.

Figure 11–7 Using sheet metal for a template.

I worked in one shop where the whole place ran on sheet metal templates (Figure 11-8). They were all numbered and stored in a special rack. Everything you needed to do the job could be found directly on the template. Hole locations are punched through the small holes and bend locations are marked in the notches. One of the best features of a template is that it is a physical representation and gage of the part, not a scaled picture or facsimile that requires mental interpretation to determine if it's correct.

Tape the center point for dividers on work that will need special finishing, or extreme shaping or forming. An alternate method is to use a sacrificial plate—put your layout and center punch marks on it (Fig 11-9). The center divot from your dividers or scriber can be a point of failure and is a pain to polish out on high finishes.

Figure 11–8 Identifying sheet metal templates.

When doing your bend layout, it is customary to indicate the direction of the bend so mistakes aren't made later at the brake. Only scribe or punch mark the inside or compression side of the bend. Bends indicated as "down 90" should be marked with felt pen only on the outside or tension side as a reminder which direction to bend. Sometimes the sheets are too large to easily flip over to properly mark the inside. Marking on the inside prevents tearing or cracking out of the scribe or punch marks (Fig 11-10).

Figure 11–9 Using a sacrificial plate.

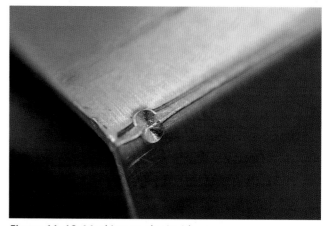

Figure 11–10 Marking on the inside.

11.2 Blank Length Calculations

Blank length calculations are one of the deep mysteries surrounding sheet metal work. The ability to mathematically predict in every situation the amount of material used during the forming of the material is one the cornerstones of sheet metal work. This little bit of mystery material is called the "Bend Allowance." It is also the cause of much frustration.

I would like to set the record straight if possible. The bend allowance is not something you can predict with absolute accuracy—period.

There are so many factors involved in calculating bend allowances that affect the final outcome of the bend that it should always be though of as an approximation as opposed to an absolute. If the sheet metal work is toleranced properly, the difference between what can be predicted and what can be achieved is the tolerance.

For the majority of sheet metal work, you can get fairly accurate results using calculations and bend allowance formulas. For most people in small shops with limited tooling, the question is how to get to get good results in a variety of material types and thicknesses. There is considerable difference in the forming results switching from .048 cold rolled and 1/8 thick soft copper.

What I am heading for here is the statement that if accurate work or extremely tight tolerances are required, then test bends will be required. There is no getting around this. Material variation alone will account for a large part of the uncertainly in predicting bend allowances.

So what we are left with are systems of blank length calculation that have different levels of accuracy, repeatability, and complexity. Repeatability is what we really care about in bend allowance calculations. The ability to reproduce your forming results, day after day and material lot to material lot, is where the rubber meets the road.

Figure 11-11 shows a graphic image of some typical bend configurations you might encounter. When we bend sheet metal, there are two questions we must address in order to produce a given bend or series of bends: 1) What length of material do I need to cut to make the part I want to make? 2) Where exactly do I place the bends on the material blank I just cut? For this explanation, I am using normal bends produced by common tooling for the example material thickness.

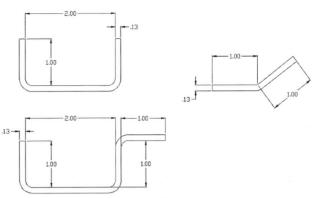

Figure 11–11a Basic bend configurations.

The first blank length calculation method is the easiest. The bends don't take much material; therefore, the assumption in the calculation is they are sharp. What sharp means is no radius is accounted for in the math. The calculation is quite simple and for a large majority of the work has more than enough accuracy. To calculate the blank length, we simply add the bend to bend lengths. In other words, we add the lengths on the inside of the shape. For the upper left example in Figure 11-11a, the blank length is

$$1.00 + 2.00 + 1.00 = 4.00$$

This is the length of sheet metal we cut to produce this channel. Accuracy in this example is easily +/− .06.

For the lower left shape in Figure 11-11a, we still add the lengths between the bends. One of the bends is on the opposite side from the two others, but as long as we add the bent-to-bend lengths, we will get the proper length. Thus,

$$1.00 + 2.00 + 1.00 + 1.00 = 5.00$$

For the upper right shape in Figure 11-11a, it's dog simple:

$$1.00 + 1.00 = 2.0$$

Now we will have to locate the bends on the blank lengths we just calculated. With the simple method true to its name, we just locate the bend lines using the dimensions off the drawing. For our first example, the channel in Figure 11-11a, we simply take our 4.00 long blank and mark a line in from the end 1.00 on each end, then bend. Done deal. For the remaining examples, they would have their bend lines marked in the same way.

That was one set of examples for several bends on non-specified material other than the thickness. For more predictable results, and I mean better than +/− .06, there are more considerations put into the calculations. Table 11.1 highlights a few factors that affect the results of bent sheet metal.

Table 11.1 Factors Affecting Blank Length Calculations	
Material type and temper	• Harder materials have greater bend allowances. More material is used per bend in hard materials.
Angle of bends	• Bends of just a few degrees use less material than acute bends or bends past 90 degrees.
Thickness of material to be bent	• The blank length calculation for accurate bends takes into consideration the neutral axis of the material. The neutral axis is the theoretical line somewhere near the center of the material thickness that is neither stretched nor compressed. It does not change length during forming, unlike the inside and outside of each bend. On thicker material, this becomes a significant addition to the math.
Type of forming tooling	• The tooling used to make the bends has an effect on bend allowances. What is calculated for bend allowances with one set of tooling does not apply to another.

You can see now why there is so much confusion around this subject of bend allowance. Everybody has their own way of doing it based on personal experience and how they were taught.

Computer-controlled press brake manufacturers have incorporated their own methods for determining bend allowances and built them into the controls of the machine.

As I said earlier, these calculations are not an exact science so they build some adjustability into the machine controls that allows the operators to adjust for the exact conditions they are experiencing.

So the next time someone tells you bend allowance calculations are an exact science, ask them why they have this adjustability in the machines. Milling machines are not adjustable like this. They cannot be operator adjusted to cut .001 one day and .010 the next.

The next method for figuring bend allowances and blank lengths is considerably more accurate and reproducible. The first step is to take the material hardness into consideration. This is the first step because the hardness of the material determines approximately where the neutral axis lies, which affects its overall length.

All the neutral axis numbers are located from the inside surface of the material or the inside radius. (Figure 11-11b). Table 11.2 shows three numbers needed to begin.

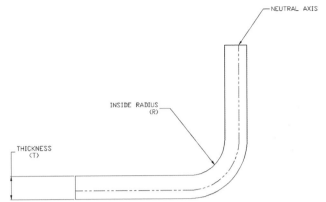

Figure 11–11b Considering material hardness.

Table 11.2 Neutral Axis Numbers	
Soft materials	• For copper, soft aluminum, lead, and gold, the neutral axis is approximately .55 of the material thickness from the inside surface—in this case, a tiny bit past halfway. Another way to write this is .55T.
Medium half hard materials	• Neutral axis = .64T. These include materials like soft annealed or killed steel, non heat treated aluminum, half hard copper, and brass.
Hard materials	• Neutral axis = .71T, and includes materials like 304/316 Stainless, cold rolled steel, phosphor bronze, and spring temper steel and stainless.

Now for the next part of the problem: the inside radius. Inside radius is calculated either from the bend produced by standard tooling or from the part drawing. Here's the rub. Now that the designer has gone and specified a radius, you are obligated at least to try to produce based on that radius. Hopefully the designer has specified the radius produced from normal tooling and not something too much larger or smaller. We're talking vanilla-flavored, 90-degree bends here. Any oddball stuff has to be worked out anyway.

The issue with the inside radius is that it may require special tools to produce. The sheet metal operator has a limited range of radii that can be produced from the normal tools for the material thickness and angle of bend.

For normal standard bends in Vee-type dies, you calculate the inside radius by starting with the material thickness and finding the correct tooling. Once the tooling has been selected, the inside radius can be approximated, which will eventually lead us into the blank length calculation. This all sounds much harder than it really is, so don't worry. That simple method is starting to look pretty good right about now.

So we have three possible conditions for the inside radius and their relative ease of producing.

1. Sharp. No calculation required for bend allowance. Easy.
2. From standard tooling. Pretty easy.
3. Non-standard specified radii. Easy-to-difficult.

In a nutshell, the inside radius from standard tooling is a percentage of the width of a Vee die 8 times the material thickness for material up to .188 thick. Putting that sentence into equation form we get,

$$R = (8T).156$$

For example, if the material is .125 thick, then

$$R = 8 \times .125 \times .156 = .156 \text{ inside radius}$$

Following through, the formula for bend allowance for a 90-degree bend is:

Bend allowance (BA) = (Neutral axis) × (Thickness) + (1.57 × Radius)

For example, for .125 cold rolled steel:

$$(.71) \times .125 + (1.57 \times .156) = .3337 \text{ Bend allowance}$$

This means each 90-degree bend uses .3337 inches of material per bend. The next step is to add the bend allowances to the straight sections of material to arrive finally at our official blank length.

Because we know the inside radius of the example is .156, we subtract that from the inside lengths from Figure 11-11. Adding it all together,

$$(1.00 - .156) + (2.00 - .313) + (1.00 - .156) + (.3337 \times 2) = 4.043$$

Applying this formula to our original example, we get a blank length of 4.0434 (Figure 11-11C).

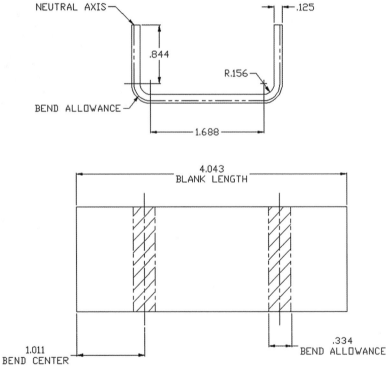

Figure 11–11c Applying the formula.

All of these calculations, just to figure the blank length, seems like a lot of work. It is, but if you want to do accurate work and have reproducible results, you at least have to learn how to do the math. All of this illustrates the theoretical method for blank length math.

I'm sure you noticed, but the blank length from the "accurate" method is only .043 longer than the length we figured using the fast method. The question is: how close do you need your bends to be? With all that calculating, these are still just numbers. Until you make a part, you are only living in the theoretical world. You will be happy to hear that there is an easer way to calculate blank lengths.

As I pointed out, the accurate blank is only .043 longer than the fast blank. Why not somehow combine these two and have an easier way of getting accurate results. You can combine them and not be stuck in the theoretical world. The way we handle it on the shop floor is to make test bends.

Test bends are the only real way of obtaining super accurate results anyway, so why not start there. If you just want the bend allowance, cut a blank (of arbitrary length) of the exact material you intend to use for your part and make a bend. Be sure to measure this test piece before you make your bend. After forming, measure it again and determine the actual loss for one bend. If you measure accurately you now have the best real-world non-theoretical example possible. The key things to remember using the test bend method are the material must come from the same lot and sheet as your part. The only other factor of significance with the test strip method is the test strip should have the same length of bend as the real part will. If the part has three inches of bend, then the test strip should have three inches also.

Charts of bend allowances for different materials are readily available to sheet metal workers.

Machinery's Handbook has an excellent section on this very subject with look-up tables for allowances. The bottom line is that most of the time you can skip the heavy math and either make test bends (recommended) or use look-up tables and still have to make test bends. The choice seems simple enough to me.

Notching patterns and angle guides. We use a simple template to set up the power notcher for repeat cuts (Figure 11-12). Figure 11-13 shows corner notches for cabinets with recessed doors.

Figure 11–12 Setting up the power notcher.

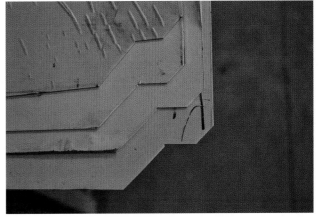
Figure 11–13 Corner notches.

Figure 11–14 A mitered corner.

This special notch allows the forming of a mitered corner (Figure 11-14).

For sheet metal layout of average size, you will need a set of large dividers and trammels (Figure 11-15). Be sure to buy some that can hold a regular pencil. These great dividers are expensive, but will sweep an 18-inch circle with a scribe point or a pencil. The bar type is superior to the dividers made from flat stock. The flat material twists when you lay into it trying to get a decent scribe going.

The little home-made device seen in Figure 11-16 clamps a faithful Sharpie marker in one leg of the dividers or trammel . Figure 11-17 shows another home-made tool that extends the range of these dividers even farther.

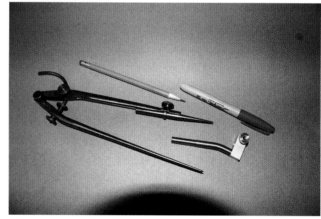

Figure 11–15 Large dividers and trammels.

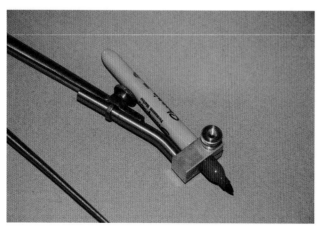

Figure 11–16 Clamping a Sharpie marker.

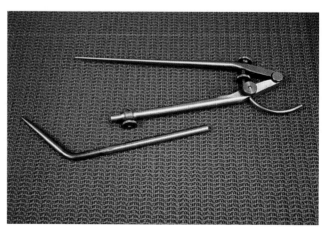

Figure 11–17 Extending the range of the dividers.

11.3 Patterns

Pattern paper is typically used for initially developing shapes other than very simple patterns. Using paper has several advantages, particularly when the pattern will be used several times or the metal that the pattern will be transferred to is expensive, such as copper, silver, or stainless steel. For sheet metal pattern drafting, a medium-to-heavyweight paper is used (Figure 11-18), if the pattern will be used in the shop or the paper pattern will actually be formed to check forming, tooling, and finished sizes.

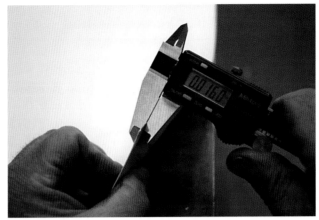

Figure 11–18 Sheet metal pattern drafting.

It's acceptable to lay out the pattern directly on the material, but a paper pattern eliminates all the construction scribe lines and arc centers normally required to develop a sheet metal pattern.

If a more durable pattern is required for marking off many transitions, light gauge sheet metal can be used. We use 26ga galvanized sheet for durable patterns. It is soft and easy to cut and the plating keeps the patterns from rusting. The original pattern is typically still done on paper. If graphic methods are being used, a sheet or sheets larger than the actual metal blank are required.

Pattern paper should be tough and fairly thick, with enough width to make an entire pattern without tacking together smaller sheets. At some of the home supply centers, red paper rolls with decent weight can be purchased. They are used to protect floors during construction. This paper is suitable for sheet metal patterns. It has enough thickness that scribe lines can be easily transferred to the metal blanks without the paper moving or the scriber slipping off the edge. Our rolls are 48-inches wide. This paper has quite a few other uses in the shop for protecting surfaces and parts

I like to secure the paper directly to my workbench with spring hand clamps (Figure 11-19). Duct tape works also but doesn't allow you to re-position the paper as easily.

Your trammels should have a wide adjustment range for the size bar they fit. They should easily hold a pencil. Fine adjustments are made by turning the eccentrically ground points. The type in Figure 11-20 fits a wide range of bar sizes, including rounds. For very long radii and arcs layouts, spin off some MIG wire and use that for your bar.

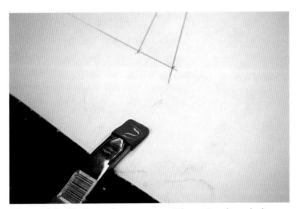

Figure 11–19 Securing paper with spring hand clamps.

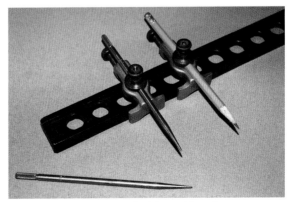

Figure 11–20 A trammel with a wide adjustment range.

11.4 The "Yank" Method

The old sheet metal crustacean I worked with taught me a neat method for transition development. From New Zealand, to him all Americans were "Yanks" and his particular method the "Yank method." It was meant as derogatory because it is not considered a "pure" method of pattern development when compared to more traditional graphical methods. I find it funny that a Kiwi hiding in Yank-land had to teach his Yank method to an American. However impure the method was, it didn't stop him from using it.

Also called the short or roll method, it uses a three-dimensional physical model of the transition to develop the flat pattern for the sheet metal. It involves connecting two ends of almost any transition together in their desired relationships, angle, distance, shape, etc., and combining them on a common axis. This model is either rolled directly on the sheet metal or rolled on pattern paper and re-used as many times as necessary. The path of the end shapes are traced as it rolls, creating the proper stretch-out and angular relationships in the flat pattern. The beauty and advantage of this method is easy to see if you have ever had to measure an odd transition in the field and get accurate results.

Figure 11-21 shows a table mockup of a typical transition you might encounter in the field. (The end shapes are irrelevant for the demonstration.) Here, we will do a simple rectangle-to-round (Fig 11-22). This set of conditions is very hard to measure and do a decent job of it. I was lucky to set this up on my welding table, which happens to have a rectangular hole in the center. Try doing that under a machine that happens to be running while you're trying to take decent measurements. This method works with almost any shape of end terminations.

Transitions from diamonds to ellipses are possible and just about anything else you can think of. No measuring is necessary to develop the pattern. The first step is to cut accurate filler plates for

Figure 11–21 Table mockup for a typical transition.

each of the ends. They should fit the openings in the same way you want the transition to fit. If the transition will fit the outside, then the plate should be that exact diameter.

Figure 11–22 Preparing for a rectangle to round.

Figure 11-23 shows the physical "Yank" mockup of the transition I am demonstrating. The two ends are locked in their exact proper relationship by the welded rods and not subject to eyeballing or sloppy measuring errors.

The mockup should have enough strength to be handled without the end moving or breaking off. I like to use 1/4-diameter rod because it's cheap and easy to cut with a small pair of bolt cutters in the field.

I always flop the mockup around through a complete revolution just to get an idea how big of a piece of paper I will need to trace the entire pattern. The example in Figure 11-24 is asymmetric so we will need the full revolution of the pattern.

For symmetrical transitions, you can just develop half if you prefer.

Trace along the edges of the mockup as it is slowly rolled through a complete 360-degree rotation (Figure 11-25). Mark the end points of the rectangle when the edge sits flat on the paper.

I also like to mark the circle at the exact spot it is touching the paper when the straight edges of the rectangle are resting on the paper. This gives me an idea where the rounding bends will be placed. It's sometimes tricky to determine where the bends will be placed; therefore, it's a good idea to work it out on pattern paper before you switch to sheet metal. If the mockup slips or is bumped while you are rolling it, you should start over.

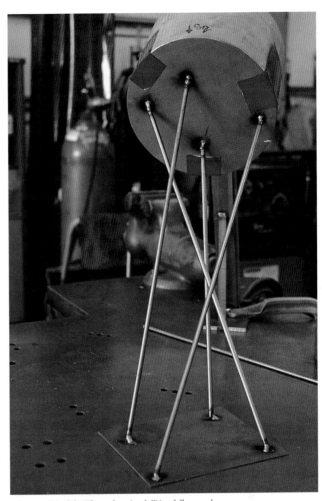

Figure 11–23 The physical "Yank" mock-up.

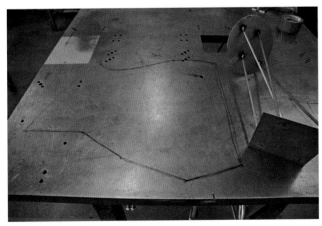

Figure 11–24 An asymmetric example.

Figure 11–25 Tracing along the edges.

In Figure 11-26, the mockup has been rolled through a complete revolution and the paper pattern developed. I have marked where the rounding bends will start and end. Remember, this is the inside of the transition. I have accidentally bent a couple inside out because I was so anxious to see the transition formed up. In Figure 11-27, you see the paper pattern fits the test setup pretty well. Paper doesn't hold its shape that well, but it gives an excellent reading on the success of your pattern development.

For the kinds of transition forming jobs, the box and pan brake is a wonderful tool (Figure 11-28). It is much quicker to set up than the press brake

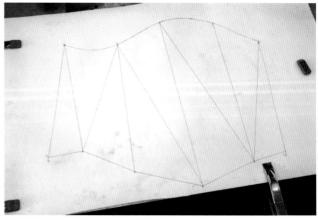

Figure 11–26 A completed revolution.

and makes these types of multiple rounding bends with a single intersection with ease. The top clamping beam is set back further than the normal material thickness to more like 1 1/2 – 2 material thicknesses.

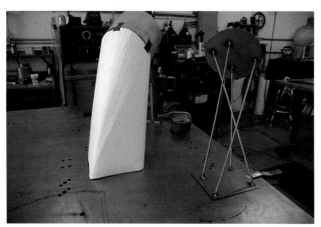

Figure 11–27 Fitting the pattern to the setup.

Figure 11–28 A box and pan brake.

Box and Pan Brakes. Box and pan brakes are versatile machines, if you know a few tricks. Unlike press brakes, they lack dedicated back gages for repeat forming work. A simple back gage can be bent up from a piece of sheet metal thinner than the part you intend to form. You can see it in Figure 11-29 under the finger. These back gages are clamped to the machine frame from the rear. Box and pan brakes are faster to set up than press brakes for a few odd brackets or multiple angle bends. Box and pan brakes are my favorite machine for bending springs clips and other small multi-angle fabrications because of setup ease.

Figure 11–29 A simple back gage.

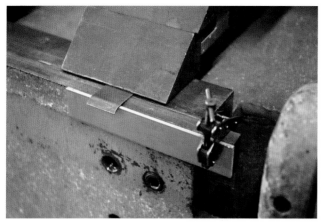

Figure 11–30 Indexing a reference bar.

Figure 11–31 Some tough snips!

Multiple bends. For multiple bends, you may find that you need to gage off the front apron. This can be done several ways. One way that saves on layout for multiple parts is to index off a reference bar or stop on the front of the machine (Figure 11-30). The box and pan brake has the ability to make opposite bends quite close together without the special tooling that would be required in a press brake.

Snipping. For those painful snipping jobs, you can get a better purchase on your snips if you clamp them in the vise. A short length of tubing completes the setup for making short work of an otherwise painful operation. The snips shown in Figure 11-31 are 25 years old and made by Bahco tool. These are some of the toughest snips I have ever used.

Corners. When forming pans and trays in the sheet metal shop, provide a corner relief hole (Figure 11-32). The thicker the material, the more important this becomes. It's easy to remember the size of the relief hole because its diameter is twice the material thickness. Use this corner detail if the formed corner will be welded. The corner will come together snugly, allowing for a nice clean seal weld.

A typical problem with common trays and pans is oil canning across the large flat surface (Figure 11-33). There are a few causes of this annoying easy-to-cure problem. The most common cause of oil canning is the corner welding.

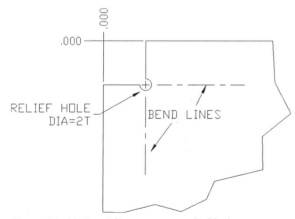

Figure 11–32 Providing a corner relief hole.

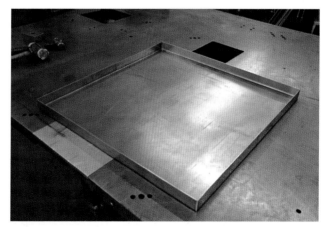

Figure 11–33 Oil canning is a typical problem.

Figure 11–34 Backing up the corner weld.

Figure 11–35 Peening the weld to stretch it.

If the corners are to be welded, be sure the flanges fit closely so minimum heat input is required. Back up the corner weld with a heat sink block of copper or aluminum (Figure 11-34). This will draw off some of the distorting heat from the joint. If you still have problems, you can use a hammer and dolly in the corner to peen the weld to stretch it a little (Figure 11-35). If your tray or pan was flat before you welded the corners, chances are the welding caused the problem.

Use a corner notch detail like the one in Figure 11-36 for formed trays and pans when the corner will not be welded. Using this relief pattern, the corners will not interfere with each other during forming. This makes for a nice straight flange without bulging corners.

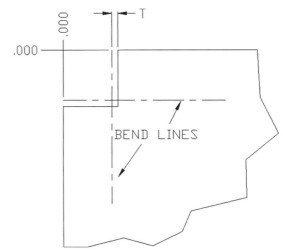

Figure 11–36 A corner notch detail.

For non-welded production trays or covers, the entire corner can be lopped off at a 45-degree angle for super simple corner preparation (Figure 11-37). Be sure to clip the corner back far enough to clear the press brake tooling (Figure 11-38). Use this corner detail so you don't have to make up a special length top punch, thereby saving setup time.

Figure 11–37 Lopping off the corner.

Figure 11–38 Clearing the press brake tooling.

Figure 11–39 A corner notch.

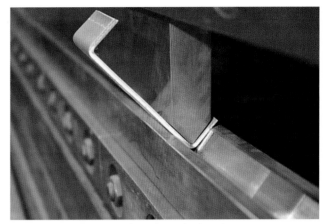

Figure 11–40 The inside radius of a bend.

The layout for the corner notch in Figure 11-39 is the flange height, plus one half the width of the top punch plus 1/16. This is the distance from the corner of the sheet for the 45-degree clip.

Bend radius. The inside radius of a bend formed in the press brake is dictated by the width of the lower die opening (Figure 11-40). Many people mistakenly believe that the upper die or punch nose radius controls the formed radius.

This is not the case, as you can see in Figure 11-41. The only difference in the two bends was the lower die Vee width. If you want to change the inside radius, change the lower Vee die width.

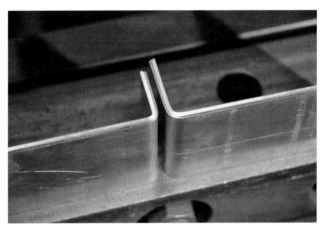

Figure 11–41 Comparing the lower Vee die width.

The upper punch nose radius should be 20% less than the calculated theoretical inside radius for the desired bend. Truth be told, the inside curvature of a formed bend is not even round. It's actually elliptical in form. This is just one reason the blank length calculation formulas are really only theoretical approximations and do not always produce accurate results.

The only difference in these two air bends was the width of the lower die opening (Figure 11-42). The smaller radius was formed in a Vee die that was 1.125 and the larger radius was a 2.00 Vee opening. Upper punch was the same for both bends.

Test bends. If your bends must be super accurate, you will have to make test bends. There is no getting around this except pure luck. With practice and consistent materials, you can get pretty

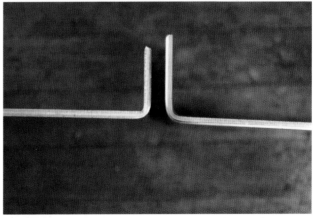

Figure 11–42 Comparing the width of the lower die opening.

close, but there is no substitute for test bends in the actual production material.

Bend radius formula. Standard air bending Vee dies want to have an opening eight times the material thickness. This ratio increases for materials over 3/8 thick where a die opening of ten times the material thickness is generally used. For instance, 1/8 thick material would use a 1-inch Vee opening lower die. The basic formula for calculating the approximate inside bend radius in steel is,

$$(MT8) \times .156$$

that is, material thickness multiplied by 8 multiplied by 5/32. This calculation gives us an inside radius of 5/32 for 1/8-thick material formed in standard air bending dies. This formula is an approximation for medium hardness materials like cold or hot rolled steel. Materials that are harder or softer than steel will not conform to this rule of thumb. Harder, stiffer materials will have larger radii and softer materials will have much smaller radii.

For some reason, many designers believe it's important to specify the bend radius for a sheet metal or plate forming operation even when it has no impact on the design (Fig 11-43). The only constructive advice I would offer is if the bend radius is not important, leave it off the drawing.

An alternative would be to apply a liberal tolerance to any specified bend radius, or note the radius as, "From standard tooling." Many hours of frustration can be avoided if you design your sheet metal bends around what your standard tooling produces for a given material. Avoid over-specifying these values if you need to send work to an outside shop.

Change bends. You can change the internal radius after forming by over-bending the angle and then back bending back to 90 degrees (Fig 11-44).

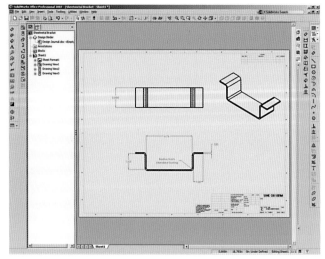

Figure 11–43 Specifying the bend radius.

The material is pulled a little further and forms a slightly smaller radius. Because of this little extra forming, the material is a little stiffer.

Therefore, when the back bending operation takes place, the net gain is a slightly smaller bend radius—a handy trick when confronted by a bend radius with a tolerance on a drawing. It's a small radius change, but sometimes necessary.

Figure 11–44 Changing the internal radius.

Figure 11–45 The channel dimension after two bends.

Figure 11–46 Changing the overall dimension.

This same trick can be applied to adjusting the distance between two bends. It is seen in bracket bending involving a closely-controlled distance between legs. This distance can be adjusted slightly by over-bending a few degrees and then back bending to vary the overall width by small amounts. Figure 11-45 shows the channel dimension after the first two bends. In Figure 11-46, the overall dimension has changed after re-striking the flanges and slightly over-bending the angle beyond 90 degrees.

Figure 11-47 shows the channel after back bending the flanges and returning them to 90 degrees. The net change in the overall dimension was .010 inches. More change can be gained in heavier

Figure 11–47 Further changes to the overall dimension.

materials and more severe over-bending. Experiment with this trick the next time you have one piece of material and you just formed the legs a little bit out of tolerance.

Bend positions can be altered slightly by pushing (Fig 11-48) or pulling (Fig 11-49) on the part as the dies close on a re-striking. Offsetting the lower die has a similar effect but is not always possible or practical on every machine. You may have to correct for slight over-bending. But if you only have one shot at your forming, this can be a useful trick to adjust a very small amount. Beware: this can also change your flange height at the same time.

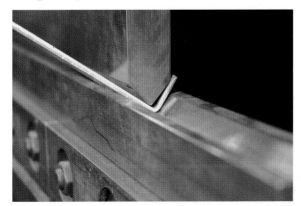

Figure 11–48 Altering bend positions by pushing.

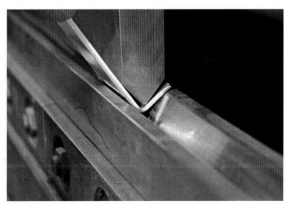

Figure 11–49 Altering bend positions by pulling.

Figure 11–50 Rough setting the depth.

Figure 11–51 Distortion that makes for sloppy fitup.

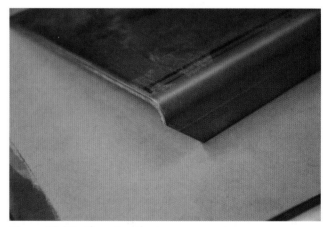

Figure 11–52 Changing the taper amount.

Forming work. A lot of forming work I learned on used expensive, thick materials. Cutting another piece of $10-per-inch flat bar or using up miles for extra test bends was not acceptable. You find a way to make small corrections for the inevitable errors in blank length and bend allowance figuring.

You can quickly rough set the depth of a set of air bending dies for a 90-degree bend by setting the bottom position with a piece of the same material to be formed (Fig 11-50).

Jog the ram down until the scrap part is almost clamped. This will be pretty close to a 90-degree bend. When forming tapered ends, keep the end of the taper away from the intersection of the bend radius. You can see in Figure 11-51 that when the taper intersects the bend line, there is a distortion which makes for sloppy fitup later on. If you increase or decrease the taper amount to move it away from the bend radius, you will get better results (Figure 11-52).

An easy way to add strength and stiffness to a bracket is to form gussets directly into the brackets (Figure 11-53). These are created by spacing the bottom and top dies, then inserting a flat bar between the lower die (Figure 11-54). The depth of the gusset is controlled by how much it protrudes into the forming area.

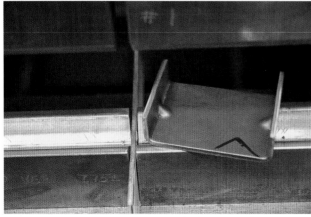

Figure 11–53 Forming gussets directly into the brackets.

Figure 11–54 Spacing the dies and inserting a flat bar.

Figure 11–55 A back bending technique.

Figure 11–56 Putting a mild bend in the center.

A word of caution: go easy. A material thickness into the forming area is all that's required. The forming bar should be rounded and smooth to prevent shearing the part and the upper die spacing needs a little extra room for the metal to expand outward along the bend. If you're forming a bunch of brackets that need a little extra stiffness, this is a great trick.

Deep channels. One trick that's handy for forming deep channels and other forms is to use a back bending technique (Figure 11-55). This is not the first choice on the list for production bending, but it can get you out of a jam for a couple of pieces when you don't have the specialized tooling to produce them.

The first step is to put a mild bend exactly in the center between the two legs (Figure 11-56). This bend should be something like 15–20 degrees, depending on the length of the legs. The 90-degree legs are then formed, making a W shape (Figure 11-57). The first bend allows the 90-degree bends to clear the machine.

The last step is to carefully align on the exact center between the two legs again, then back bend the W flat again (Figure 11-58). I find it helpful to mark the centerline on both sides of the blank before I start.

Figure 11–57 Making a W shape.

Figure 11–58 Aligning and bending back.

Figure 11–59 Shop-made gooseneck dies.

Figure 11–60 Finding the correct die.

Gooseneck dies. Here is a shot of some shop-made gooseneck dies (Figure 11-59). You just never seem to have just the correct die for some of those tightly-formed channels (Figure 11-60). These were made from a couple of heavy angles in the fabrication shop in about an hour.

Figure 11–61 Sandwiching perforated sheets.

Figure 11–62 Sheets that already have holes and cutouts.

Perforated sheets. Sandwich perforated sheets that have a linear pattern or holes that will distort when rolling (Figure 11-61). This method also works well for sheets that already have holes or cutouts in them (Figure 11-62). Sandwich the work piece between two thinner sheets, then roll normally. The results will be a facet-free cylinder (Figure 11-63). The inner sandwich sheet needs to be a little shorter than the part sheet; otherwise, there will be an overlap problem as the cylinder closes.

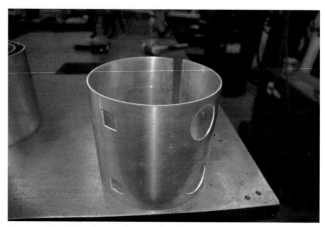

Figure 11–63 A facet-free cylinder.

Figure 11-64 A window punch.

Figure 11-65 Forming the legs closely together.

Window punch. Here is a special punch for the press brake, sometimes called a window punch (Figure 11-64). This cutout allows the legs of a channel to be formed closely together (Figure 11-65). It is very limited on applied tonnage, but handy for those impossible bends that seem to crop up. This punch was shop-made for a special shackle-forming job.

Large cuts. When making large cuts on the shear, a simple trick to save your back is to slip the width of the sheet on top of the back gage to make your roughing cuts (Figure 11-66). This keeps the heavy drop from falling down into the bottom of the drop chute behind the shear. You can now go around the back of the shear and slip the large drop back up onto the deck of the shear for accurate squaring. Believe me, if you have to shear stacks of heavy material all day long, this will save your spine some serious abuse and probably several trips to the back cracker's couch. This trick may not work on all shears because of the height of the back gage and the width between the housings. Set the gage at approximately half of the length of the drop. Doing so allows it to balance on the back gage and let you slip it back onto the shear deck.

Repetitive cuts. For repetitive cuts, front stops keep the sheet on the operator side of the machine for faster loading and unloading for large pieces (Figure 11-67). Once set, these are very accurate; they're not affected by sheet droop like the back gage. A quick way to set the front stops accurately is to shear a strip the exact length you need. Then jog the shear so the top blade is just crossing the bottom right in line with the front stop. You can now butt the strip against the blade and run the front stop into position. If you have a mechanical shear, you can cut individual strips of scrap material and measure the length to adjust the stops accurately. This trick, combined with the other trick shown in Figure 11-66 using the back gage to support large drops, can save you a lot of hard work.

Figure 11-66 Making large cuts on the shear.

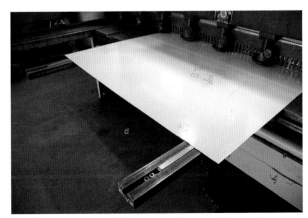

Figure 11-67 Using front stops for repetitive cuts.

Figure 11–68 Stripping when shearing small pieces.

Figure 11–69 Blanking when shearing small pieces.

Small pieces. Always strip (Figure 11-68), then blank (Figure 11-69) when shearing small pieces. The order of shearing has a great affect on accuracy and speed. Long narrow strips are hard to handle and cut accurately. The narrower and longer the strip, typically the more twist it has.

Tacking parts. Here is a simple little trick to get your sheet metal parts tacked together. Sometimes it's a chore to hold both the part and the TIG torch, and then on top of that step on the pedal to get a part tacked. If you put one part out of alignment in one axis (Figure 11-70), you just align the corner of the joint for tacking; it makes this juggling act a little simpler. In this case, the sheet metal fitup is inside corner to inside corner. All we need to think about is keeping a little downward pressure and the parts will stay in alignment.

After you get a small fusion tack (Figure 11-71), the top sheet can be rotated carefully into correct alignment for the next tack. All this is done one handed.

Figure 11–70 Putting a part out of alignment.

Figure 11–71 A small fusion tack.

11.5 Forming and Layout of Cones

The cone is one of the basic shapes encountered in sheet metal. Years ago, somebody gave me a little sheet metal handbook that had a great little method for laying out truncated cones, which is how all fabricated cones are made. This is the method I use exclusively now.

The diagram in Figure 11-72 shows cone dimensions and naming conventions. Eventually you won't need the diagram. You may even make a cool little spreadsheet that you plug your numbers into to get the output.

Figure 11-72 shows the actual flat pattern. The secret to the method's accuracy is the chord dimension (Figure 11-73). Because we are using the chord of the arc segment, we bypass the need to accurately measure the angle and the circumference. On small cones, this is usually not a big deal. On large cones, it's difficult to get accurate results by using the angle to determine the pattern.

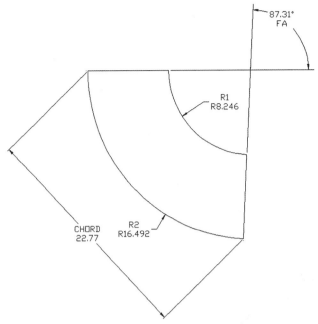

Figure 11–73 Key elements for frustum development.

The problem is the arc length defined by the angle becomes the circumference for the base diameter and frustum diameter. If these are not the correct length, the cone diameters will be wrong. When measuring a large arc with an angular measuring tool, it's easy to miss by a quarter or half degree, affecting the finish diameter. It's always easier to measure a straight line like the chord than an accurate angle. Therefore, we just use the chord dimension to locate the end points on the pattern.

The second benefit of this method is that as soon as you have the chord dimension, you know how large a piece of material will be required. Note: if the flat angle is larger than 180 degrees, the chord dimension can't be used for sizing your starting piece of material. Instead, double the large radius dimension blank length and add a little bit for trimming.

Working through the example in Figures 11-72 and 11-73:

Base Diameter = 8.00   Height = 8.00
Frustum Diameter = 4.00

SH = Slant Height   FA = Flat Angle
CH = Chord

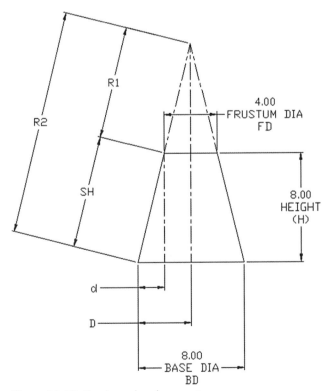

Figure 11–72 Frustum development.

Try it out yourself with some little paper cones at your desk. You will be impressed with the accuracy of this method and the simplicity once you have done a few. I just noticed that the difference between the two radii is exactly half. I didn't plan that when I set up the example cone; it must just be the math gods working their mojo.

When you form conical shapes in the power rolls, the ends of the cone blank need to travel through the rolls at different rates.

Depending on the difference between the two diameters of the cone, this can be difficult to control while rolling. One little helpful trick is to use a stalling or holdback pin. The purpose of the pin is to retard the smaller end of the cone so it travels through the rolls at the same rate as the larger end. The block and pin shown in Figure 11-74 were fabricated to fit our small pyramid power rolls. The pin is spring loaded so it can be pushed down into the block as the top roll is lowered.

Figure 11-75 illustrates using a holdback pin to stall the small end of a cone formed in the power rolls. What this holdback device looks like will depend on the configuration of the rolls you have. For initial pinch rolls, a simple length of angle

Examples from Figures 11-72 and 11-73	
Formula	**Application**
$SH = H2 + d2$	$SH = 82 + 22 = 8.246$
$FA = \dfrac{d360}{SH}$	$FA = \dfrac{2(360)}{8.2462} = 87.3129°$
$R2 = \dfrac{SHD}{d}$	$R2 = \dfrac{(8.2462)4}{2} = 16.4924$
$R1 = R2 - SH$	$R1 = 16.4924 - 8.2462 = 8.2462$
$CH = (\sin)1/2(FA(2R))$	$CH = .6903(32.985) = 22.7706$

Figure 11–74 This block and pin fit small pyramid power rolls.

iron, with a tongue the same thickness as your material and the Vee pointed into the pinch, will work for holding back the small end of a cone. The second desirable feature of this holdback device is it keeps the cone blank away from the housings of the rolls during rolling (Figure 11-76). Without the holdback device, the blank can drift and interfere with the machine.

Figure 11–75 A holdback pin.

Figure 11–76 Keeping the cone blank away from the housings.

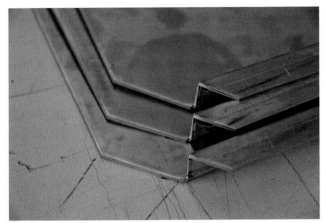

Figure 11–77 Corner notches for cabinets with recessed doors.

Figure 11–78 All the bends have been made.

Corner notches. Here is a funny little corner notch we use to fabricate cabinets with recessed doors (Figure 11-77). Surface-mounted doors always look lousy to me so we use this method to set the door into a recess. In Figure 11-78, all the bends have been made creating a nice mitered corner with a door recess (Figure 11-79).

Figure 11–79 Creating a mitered corner with a door recess.

Rims and Edges. Here are a couple of ideas for rims and edges. Instead of doing a bunch of extra notching work on the sides of a box or tray-shaped part, it's sometimes easier to insert a separate filler piece. In the example in Figure 11-80, the flange is turned outside which means we either have to do a deep complicated notch or, in our case, add a simple filler piece.

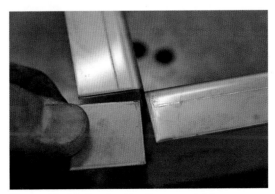

Figure 11–80 Turning the flange outside.

Figure 11–81 The filler piece is larger than the space.

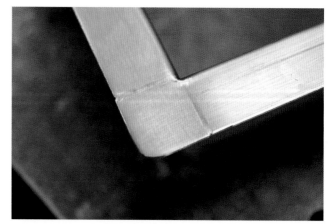

Figure 11–82 Making the flanges flush.

Make sure the filler piece is slightly larger than the space (Figure 11-81). This allows your weld to go all the way to the outside edge and still leave a little material to sand back to make the flanges flush (Figure 11-82).

The tray in Figure 11-83 has an extra flange bent on it. This makes for a very stiff part, and a tray that is easy to pick up because of the extra bend.

Filling in the corner is a little more complicated. Most folks just do a 45-degree filler piece, but I want to show a fancier rounded corner.

Figure 11-84 shows the start of the filler piece with notches. It's the flat piece that looks like a stealth bomber. After forming and fitting we end up with the part in Figure 11-85. This can now be welded in place.

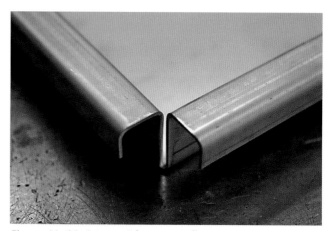

Figure 11–83 A tray with an extra flange.

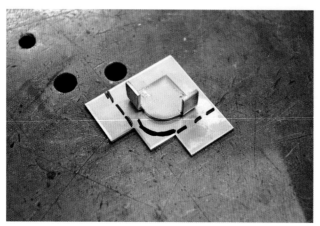

Figure 11–84 The start of a filler piece with notches.

Figure 11–85 After forming and fitting.

Figure 11–86 A small strip that follows the curve.

Figure 11–87 The finished tray corner.

The last bit is a small strip that follows the curve and makes up the curved vertical part of the filler piece (Figure 11-86).

After a little sanding and detailing, we have a nice looking tray corner (Figure 11-87). Functionally, this is the same as the 45-degree variety, but I like the look of the fancier rounded corner.

11.6 Tanks and Baffles

Many years ago my old sheet metal teacher taught me a way of installing baffles in tanks worth mentioning. Baffles are internal panels inside storage tanks and are used to control sloshing and shifting of the tank contents. Many tanks do not require baffles, but almost all tanks used in equipment that moves or is moved will be baffled. The requirements for baffles are that they are strong enough to resist the material inside the tank pushing against them, and that they slow but not block the passage of the material. For this reason, many baffles have holes or cutouts in them to allow the controlled passage and flow of the tank contents.

Figure 11–88 A typical tank baffle.

Figure 11-88 shows a typical tank baffle. The lower corners are clipped so material can pass back and forth between the two sides. In effect, a baffle creates two smaller tanks. The number of baffles required is related to the minimum size tank you would not bother to baffle.

Installing baffles in the tank body can be a little tricky. It is important for the baffle to fit the tank body properly so there are no exterior bulges or concavities, but still be easy to slip into place. Not too loose and not too tight is the right way for a baffle to fit.

Figure 11–89 A short flange bent on three sides.

Figure 11–90 The flange is partially consumed.

The secret of this method is the short flange bent on three sides of the baffle (Figure 11-89). This makes installation of the baffle much easier. The flanges act like legs and keep the baffle in position for welding. They also make the size of the baffle adjustable by changing the bend angle of the flange. The third benefit of the flange comes during welding when the flange is partially consumed in the weld as filler without damaging the tank body (Figure 11-90).

When you need to place a lid over an opening, a handy trick is to use thin supports diagonally (Figure 11-91). This trick still allows you to slide the lid around to get it tacked up without raising it too high. Use thin material so after you have a couple of good tacks on the lid, you can still slide the supports out (Figure 11-92).

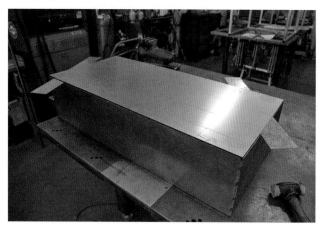

Figure 11–91 Using thin supports diagonally.

Figure 11–92 Sliding out the supports.

The Abrasion Depart- ment

12

12.1   Sanding, Grinding, and Abrading
12.2   The Good, the Bad, and the Ugly
12.3   Radius Grinding

12.1 Sanding, Grinding, and Abrading

A better title for this chapter might be "Finish Work." Many times, this work is the last step before the job is finished and proudly presented to the world. It is also often where a job or part goes from great to awful in the blink of an eye. A bad final finish is difficult to recover from without adding a lot more work. The trick is knowing when to stop and say "enough!" then taking the time to make sure the last step is just that—last!

How many times have you worked so carefully for many hours, only to botch the last step like slipping with your de-burring tool or part number marking, and really take an otherwise perfect job spiraling down the drain? It might be something as dumb as dropping the part on the hard floor or filing a small burr off that is the cause of the damage.

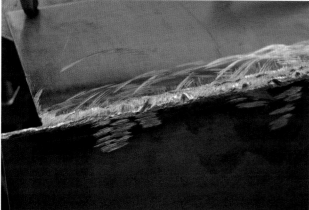

Figure 12–1 Quit while you're ahead!

Figure 12–2 Going too far.

As humans, we like to complete a task or accomplish something, but another little part of us cannot leave well enough alone when we really know better. I call this syndrome, "Just one more little thing." There is a critical juncture in every job where any further input by humans is really unnecessary and comes with an exponentially increasing risk of a screw up. Old timers have messed up enough so they recognize these forks in the road and heed the signs. They know when to get off the horse.

Figures 12-1 and 12-2 show before and after the point when somebody should have gotten off the horse….Don't try this at home, kids! No amount of additional grinding will help the part in Figure 12-1, as you can see in Figure 12-2.

Dress cutting discs. Dress abrasive cutting discs with a stone to maintain free cutting action (Figure 12-3). Using a stone is a trick that works to keep your cutting discs humming along. In general, these wheels break down quickly enough that fresh abrasives are exposed easily. The thicker the cutting disc, the better this technique works, especially when you have a broad cut in comparison to the width of the abrasive disc. Use this trick on chop saws or broad cuts in tough materials, or materials that tend to load the wheel, like stainless and aluminum. A quick touch-up with a dressing stone sharpens it in a flash. In Figure 12-4, I am using a boron nitride dressing stick called Norbide. This tough super-hard material makes a great cutting wheel dresser.

Figure 12–3 Using a stone to dress abrasive cutting discs.

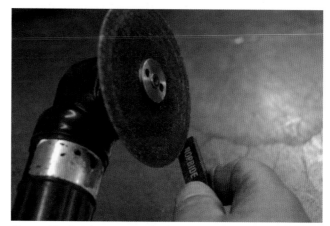

Figure 12–4 A boron nitride dressing stick (Norbide).

Figure 12–5 Using a round grinding stone.

Figure 12–6 The importance of good fitup.

Tube fitup. You can use a round grinding stone to make perfect joint fit ups in round tubing and pipe (Figure 12-5). These round grinding stones are available in several diameters. If I am fitting tubing that has an odd diameter, I would dress the round stone with a diamond in the lathe until it was a close match to the tube diameter. Good welding can only come after good fitup (Figure 12-6). This is one way to make the joints fit with a minimal gap. Use this method for those off-angle connections or the tube sizes that your notcher cannot do.

Graining nails made from stainless filler rod. When you need to add a linear grain finish to stainless steel sheet, it can sometimes be a pain to clamp the sheet so you don't get funny runoff marks or stops and starts where the clamps were. One trick I have used is what I call "Graining Nails." I like to use a piece of plywood to grain flat sheets manually. The plywood should be a little wider and a little longer than the sheet you are working on so you can clamp it securely. The graining nails look like small finishing nails and are made from snipped off pieces of stainless filler rod (Figure 12-7). We use stainless nails for stainless and steel nails for steel. Using materials that are alike prevents steel contamination of the stainless sheet from unwanted metal transfer. A few of these around the perimeter secures the sheet during graining without using clamps (Figure 12-8).

Cut Scotchbrite sheets down to smaller size. The full-size Scotchbrite hand pads are much too big for normal small blending and finishing jobs. Most of the time, you just need a little square to blend a little area. If you cut the Scotchbrite pads down into three finger-sized little pucks, you can fully use the abrasive and get many more square inches of blending out of each pad. Cut them with a crummy pair of old tin snips. Never use your good snips to cut abrasives.

Figure 12–7 Graining nails.

Figure 12–8 Securing the sheet during graining.

Grinder holders (Figure 12-9). You can tell if somebody knows how to grind by the way they lay their grinder down. The goal in good grinding is to keep the disc as flat as possible for the life of the disc. A warped disc makes for crappy grinding and jello arms at the end of the day. A better way to hang up a grinder with a sanding disc is to make a holder that protects the abrasive disc from damage (Figure 12-10).

Figure 12-11 shows the wrong way to lay a grinder or sander down. Most grinders have those nifty little black plastic bumpers on the back side. Why not use them? If you lay your grinder down like this for any length of time, the disc will warp and it will show up in your finish work. If you grind stainless steel, laying your abrasives down like this on a steel table will contaminate the abrasive and transfer the contamination to the stainless steel.

Figure 12–9 Grinder holders.

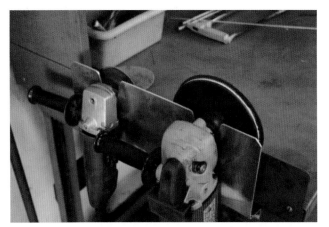

Figure 12–10 Protecting the abrasive disc from damage.

Figure 12–11 The wrong way to lay down a grinder or sander.

Sanding disc flattener. It's a real pain when you're sanding discs warped from improper storage or high humidity. Everything is fine until you open the package (Figure 12-12). Warped discs only cut on the high points, so it's important that your discs are pretty flat if you want to do decent work and get the most from your premium abrasive dollars. The simple shop-made flattener in Figure 12-13 keeps a little pressure on the discs as the humidity changes.

Figure 12–12 Warped discs.

Figure 12–13 A shop-made flattener.

Figure 12–14 Rotating the backing pad.

Another trick to extend the life of your sanding discs is to rotate them after you have used them for a while. The trick is to rotate them in relation to the backing pad. As you can see in Figure 12-14, this disc has not been cutting evenly all the way around. The spot I am pointing at has seen almost no work. When you rotate it slightly in relation to the backing pad, you can sometimes move the low point of the disc to an area on the backing pad that is higher.

Wrap abrasive around a stiff flat bar for better cutting action. The hard backing of a metal bar converts more of your elbow grease into removed metal (Figure 12-15). Thin flat-bars sneak into gaps and openings where regular sanding pads won't fit (Figure 12-16). It might be helpful to think of this idea like a fine grit file. The action is very similar. You can use a small clamp to secure the abrasive from moving.

Figure 12–15 Wrapping abrasive around a flat bar.

Figure 12–16 Using thin flat bars.

Tape abrasive paper to the surface plate to keep the wrinkles out (Figure 12-17). I hate it when you have a fresh sheet of silicon carbide paper to do a little flat lapping and the first thing that happens is you crinkle the paper and ruin it for lapping. This trick allows you to concentrate on the lapping job instead of holding the paper steady.

Figure 12–17 Keeping wrinkles out.

Change the direction of sanding to give a balanced crosshatch pattern (Figure 12-18). I call this fake surface grinding. It's not meant to deceive anybody, instead just to mimic a Blanchard ground surface. In some cases, it looks better than the standard figure eight pattern normally used. It also helps you see the scratch lines left from the previous grit abrasive paper.

Add a little light machine oil or WD-40 to make an abrasive act like a finer grit (Figure 12-19). This trick reduces the effective grit by approximately half, allowing you to do your finish with the same abrasive grit (Figure 12-20). This also works well with the ball holes mentioned elsewhere. The kerosene in the WD-40 keeps the abrasive clean and cutting.

Figure 12–18 Changing the direction of sanding.

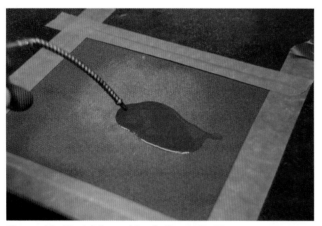

Figure 12–19 Adding a bit of oil or WD-40.

Figure 12–20 Reducing effective grit by half.

Ball hones. Internal ball hones have saved my bacon quite a few times (Figure 12-21). These cheap and easy-to-use hones have a dozen uses around any shop. They work equally well on soft and hard materials using a common cordless drill (Figure 12-22). I have even used these in a pinch to resize a commercial ball bearing bore to fit a slightly oversized shaft. They can de-burr those nasty cross-drilled holes and steps in a bore (Figure 12-23) as well as improve the surface finish an easy ten points.

Figure 12–21 Internal ball hones.

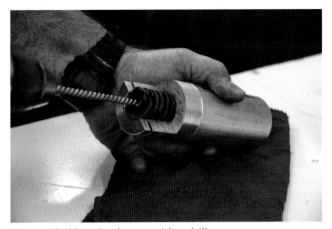

Figure 12–22 Using hones with a drill.

Figure 12–23 De-burring cross-drilled holes in a bore.

De-burr in the drill press. Here is an interesting de-burring trick. Take steel wool or, in the case of this example, bronze wool and wrap it around a piece of threaded rod in the drill press (Figure 12-24). This mild cutting action is just the ticket for many delicate de-burring jobs. The threaded rod has the right amount of tooth to retain the wool and keep it from spinning on the rod (Figure 12-25). The cutting action is similar to a non-woven abrasive, but with a greater polishing ability.

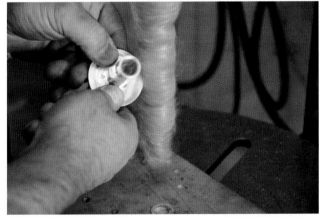

Figure 12–24 Wrapping bronze wool around threaded rod.

Figure 12–25 Cutting action similar to non-woven abrasive.

Change direction 90 degrees each time you change abrasive grits. This makes it easy to see the scratch lines from the previous grit. There is nothing worse than trying to get 100 grit lines out with 600 grit paper. I think the numbers mean the number of hours to remove the scratches...!

Figure 12–26 A kit for producing fine finishes.

Figure 12–27 A stainless flat bar.

Stainless polishing. Figure 12-26 shows one of the best kits I have ever seen for producing fine finishes on stainless steel. The kit consists of all the abrasives needed to go from weld grinding the entire way to a mirror finish. They are all correct for the incremental finishing needed to produce a high polish on stainless and many other metals. Walter Abrasives has collected all these grits into a great little kit called "Quickstep."

You may want to test a piece of stock stainless flat bar like that featured in Figure 12-27. You can see the mill finish and even a boot print where somebody walked on it!

Starting with the first flap disc in the set (Figure 12-28), each abrasive in the set is just the right amount finer for each finishing step. Be sure to change sanding direction as you work your way up into the finer grits. If you do not remove all the scratch marks from each grit, you will have to go back and remove them with a coarser grade.

Now it's hard to take a picture of a highly reflective surface, but I think the one in Figure 12-29 does the results justice. You can see the reflection of the screw of the Bessey clamp in the polished surface. One thing I have found important is to limit the heat in the part. Cool it off once in a while to keep from melting or damaging the finer abrasives in the kit. I spray water on the part and blow it off with an air hose to cool it between grits.

Figure 12–28 Starting with the first flap disc.

Figure 12–29 A highly reflective surface.

Figure 12–30 Squaring the table of the belt sander.

Figure 12–31 Replaceable front edges.

Figure 12–32 Reducing the gap.

Figure 12–33 Protecting the tiny part.— and your finger!

Figure 12–34 Marking witness lines.

Belt sander table. Square the table of the belt sander before you start (Figure 12-30). The table tends to droop downward during use if not checked. Therefore, if you have a finish squaring job that you want to look nice, then it is only a quick second to check and adjust. On this same note, the belt platens will need replacement or re-surfacing every few years to maintain a decent flat backing for nice detailing work.

In Figure 12-31, we have added replaceable front edges to the belt sander tables. This allows us to keep the gap between the belt and the table close and not have to replace the table casting every year.

When you have to detail very small parts, use a piece of flat bar to reduce the gap between the disc and the worktable (Figures 12-32). This prevents your tiny part or, worse, your finger from getting sucked into the gap and lost forever (Figure 12-33).

If you sand a bunch of parts at a particular angle, mark witness lines on the table of the belt sander to guide your roughing work (Figure 12-34). This visual cue keeps you from drifting too far from your intended angle. You can clamp a guide to the table, but that destroys one spot on the belt.

Figure 12–35 Using a thin masking plate.

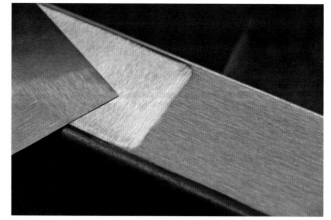

Figure 12–36 Changing the graining direction.

Masking plate. Use a thin masking plate when you have intersecting linear finish lines. This produces a crisp demarcation line between the two intersecting finishes. Be sure to use a plate of the same material to avoid contamination. In Figure 12-35, the masking plate is .010 stainless shim stock.

The long length was grained first, then the mask was used when the graining direction was changed (Figure 12-36). It's easier if you have something with a little length to it, but this demonstration gets the idea across.

Mark abrasives. Mark the materials you use your abrasives on and segregate them from the other discs (Figure 12-37). I have seen some beautiful jobs ruined by a contaminated sanding disc or even a dirty wire brush. When in doubt, use a new disc. One of the guiltiest parties in the shop with regard to contamination is the dirty little wire brush on the bench grinder or buffer (Figure 12-38). It spreads its corrosive disease to everything it touches. You may have the best intentions, but this wheel is almost always contaminated with steel, rust, and grease in the average shop.

Figure 12–37 Marking the materials.

Figure 12–38 A contaminated wire brush.

Laps and Lapping. Figure 12-39 shows a few examples of laps and lapping. These are all cylindrical laps for OD and ID work. Simple to make in the shop, they produce some fantastic results. These particular brass laps were used to lap a bearing bore (far right) and the OD of a precision gage (center). The results are controllable down to micro-inches if necessary. Here, the lapping compound was diamond paste. The laps themselves need to be accurately made, but with a little effort results that cannot be attained any other way are possible. All the laps shown in Figure 12-39 are used under power at slow speeds, approximately 100–300 RPM. and stroked axially over the length to be lapped to produce a crosshatch pattern of approx 30–45 degrees.

Dress your surface grinder wheels at a slight angle for peripheral roughing work (Figure 12-40). The edge will break down faster, exposing sharp grains for faster stock removal. Re-dress for the fine finish work. For roughing, use an aggressive depth of cut and step over as much as the wheel will tolerate without complaint. The abrasive grains need to break down to continue the cutting action. If they are babied and allowed to glaze over, all your effort just turns into heat. Abrasives should be thought of as little cutting tools, much like a lathe or milling machine cutting tools. Properly used, the chips look similar under high magnification.

A similar roughing trick works for side wheel grinding, which on a good day is a bit of a pain. Side wheel grinding is not an efficient material removal method, but is necessary some times because of setup issues or part geometry (Figure 12-41). Dress the side of your wheel as you would normally. Stop the grinder and loosen the nut. Rotate the wheel slightly in relation to the spindle. Tighten it all back up and go ahead and do your side wheel roughing. What happens is the wheel will run out a little from where it was dressed and only cut on one part of the periphery. This small edge breaks down quickly and continues to cut with the need for redressing during your roughing. Dress again and don't rotate the wheel for your finish work. This trick can give you more grinding time between dressings, which makes a difficult grinding job go faster.

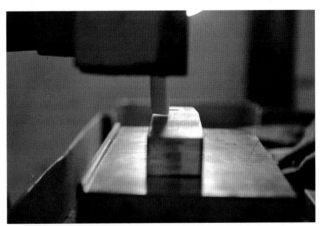
Figure 12–40 Dressing surface grinder wheels.

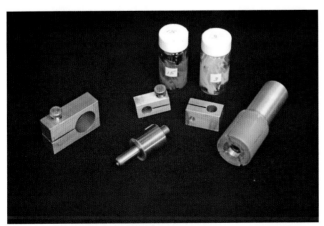
Figure 11–39 Laps and lapping.

Figure 12–41 Side wheel grinding.

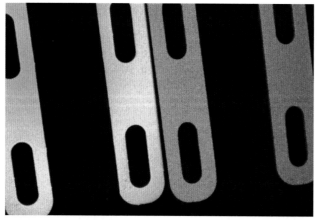

Figure 12–42 Glass bead blasting.

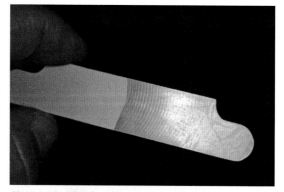

Figure 12–43 Blending tool marks.

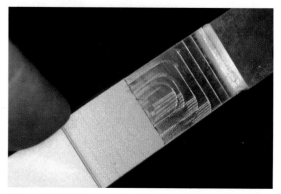

Figure 12–44 Removing tool marks.

Bead blaster finishes. Glass bead blasting can de-burr and produce finishes that are hard to produce any other way. Anything from a fine satin finish all the way to a rough surface to add grip or traction to a part can be produced with a few different grits and by varying the blasting pressures used (Figure 12-42).

Tool marks can be blended or removed with fine mesh size glass beads (Figures 12-43 and 12-44).

Beware of large sheets and surfaces. The bead blaster is like hammering with a billion little ball pein hammers; it will move material if you're not careful. The bead or sand blaster is another often overlooked source of material cross contamination. If you have been cleaning engine parts for your jeep, you probably need to replace the media if you're going to blast some medical parts.

Figure 12–45 A tea strainer trick.

Figure 12–46 A laser cut mask.

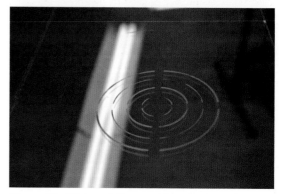

Figure 12–47 Imprinting a polycarbonate sheet.

Ever had a hard time holding onto small parts in the bead blaster? It's really a pain if you drop one in the bottom of the blaster and have to fish it out. This little tea strainer trick keeps the part under the beads and captured (Figure 12-45). Be sure the mesh opening is larger than your beads or it will take a really long time to do the job. I have shown this opened up so you can see inside. A spring closes the two halves together when in use.

Patterns and logos can be etched or imprinted on a surface by using a simple mask and abrasive blasting. The mask in Figure 12-46 was laser cut for this specific purpose to produce a pattern on glass sheet. In this example, I did a piece of polycarbonate sheet and left the factory masking paper on the plastic sheet while I blasted it (Figure 12-47).

Figure 12–48 Using carpet samples.

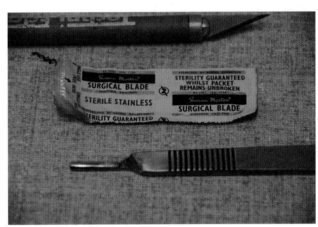
Figure 12–49 A surgical scalpel blade.

Carpet Protectors. Now you have a use for all those cheap, nicely-hemmed carpet samples from the flooring stores (Figure 12-48). They make great throw-away portable part protectors and backup surfaces for DA detailing. When they get loaded with dust, toss them out and start again.

Surgical de-burring. For fine de-burring, try a surgical scalpel blade (Figure 12-49). The sterile stainless blades seem to be considerably sharper than the plain Jane carbon steel models. If you don't like it for de-burring, you can always do a little amateur shop surgery with it!

Ceramic de-burr knife. The ceramic-bladed scraper seen in Figures 12-50 and 12-51 is superior for gummy grabby plastics like UHMW and ABS. These materials can be a challenge with clean, crisp de-burring jobs. The ceramic has just the right amount of edge sharpness to peel, without gouging softer materials. Another trick is to put soft plastics like this in the freezer if you have the time. This makes the burrs more frangible and easier to snap off cleanly.

Figure 12–50 A ceramic-bladed scraper.

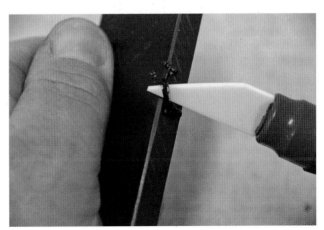
Figure 12–51 Working with gummy plastics.

Figure 12–52 Vibratory tumblers.

Figure 12–53 Media wedging.

Vibratory tumblers. Vibratory tumblers are a double-edged sword with the potential to decrease your hand de-burring time by a huge amount or destroy an otherwise simple job in a matter of minutes (Figure 12-52). The secret is all in the media and the part geometry. The deadly booby traps are media wedging and clogging (Figures 12-53 and 12-54), and material contamination, not to mention just plain losing small parts. You can lose huge amounts of time just trying to find small parts if you're not set up properly. Run sample extra parts first before you commit a large batch of critical parts. Beware of overloading larger parts that can clank together causing serious damage. It's a real pain to remove ceramic media from blind tapped holes without messing something up. Always note a proper count when you drop a batch in. There's nothing like finding that one missing part two months later when you change media.

For really small parts, you can use a sub-container inside the main bowl (Figure 12-55). This prevents tiny gray parts from becoming hopelessly lost in a sea of tiny gray media. It's also a neat trick to allow you to switch media sizes and shapes without having to dump out 300 pounds of rocks for your one little part.

Figure 12–54 Clogging.

Figure 12–55 Using a sub-container.

Figure 12–56 Masking features for protection.

You can mask features you want to protect or stay sharp while the rest of the part goes through the vibratory tumbler de-burring cycle (Figure 12-56). Plastic or rubber caps and the mesh sleeve used to protect shafting work great for masking. You can also use small silicone plugs for masking holes you don't want media to enter.

You can secure small parts to a leash of sorts to make them easier to retrieve from the tumbler (Figure 12-57). I use stainless tie wire and a large "flag" that I can spot easily when I open the lid. This cuts down on the inevitable search and count associated with media tumblers without parts discharge devices.

12.2 The Good, the Bad, and the Ugly

Little grinders are for little work. If you have any serious material to remove, do yourself a favor and get a large sander or grinder. Seven-inch seems to be the optimal size for normal-sized people. Nine-inch diameter wheels have an annoying gyroscopic effect that tends to fatigue the operator more quickly without much metal removal advantage. Seven-inch abrasives seem to be more readily stocked and available in a wide array of abrasive materials and grits.

I am a firm believer that you get what you pay for in abrasives. Cheap abrasives are just that—cheap. The longer an abrasive disc lasts before it glazes and just produces heat is the true measure of performance. For sheer metal removal and fine control, I would put a coarse sanding disc up against any hard type disc for metal removal rate any day (Figure 12-58). The only exception to that would be in an application where the amount of abrasives used per hour outweighed the labor costs for changing discs. In other words, sanding discs cut faster, but don't last as long as hard discs.

Figure 12–57 Securing parts to a leash.

Figure 12–58 A coarse sanding disc.

Figure 12–59 Advantages of hard discs.

Figure 12–60 Fracturing tough to grind material.

Hard discs can do some things sanding discs cannot do, like cut a bolt off flush with the floor or any application that needs the front edge of the abrasive wheel to remove metal (Figure 12-59). The main advantage of the hard discs is for slot or recess grinding and longevity.

Another application where hard discs are superior to sanding discs is in a situation when you have hot-rolled mill scale to remove. The hard disc fractures this tough-to-grind material (Figure 12-60) whereas a sanding disc just polishes itself into oblivion (Figure 12-61). Hard discs have their place in every shop, don't get me wrong. They last longer, but don't cut nearly as fast. At today's labor rates, which would you prefer to pay for?

Most people wait way too long to change discs. The telltale signs are excessive heat and a polished surface. The sanded surface should have a matt or satin finish to it and appear flat when the abrasive is cutting properly. If you examine the disc, you can see the polished surface of the glazed abrasive surface grains. Sometimes you can dress these out with a dressing stick to get a little more life out of the disc. I tear the disc when I throw it in the trash so the cheapskate dumpster divers don't pull them out and waste time with discs that are worn out.

Figure 12–61 Polishing tough-to-grind material.

Figure 12–62 Decking off the weld.

Figure 12–63 Light faceting passes.

12.3 Radius Grinding

A technique I call roll grinding is used to finish corners to match bend radii. Long weld seams that need to be rounded can be handled using a rolling technique with the disc sander. It takes a little practice, but once you have it mastered the results look just like a formed corner. In one shop I worked in, we did miles of seams like this. Good weld technique can minimize the amount of metal removal for the fine finishing.

The weld should first be decked off flush against the two intersecting right angle surfaces (Figure 12-62). For bumpy welds, I sometimes make a few light faceting passes to even everything out before the rolling (Figure 12-63). If you're right handed, stand with the weld to your right. Your right hand is the balance hand that supports the weight of the grinder and your left hand does the precision rolling motion. The wheel or disc should be kept as flat as possible on the weld seam. Very little pressure is used to do the actual rounding. Think of your right hand as the pivot holding a round object like a motorcycle grip or a piece of round handrail. As you sand, step slowly backward along the seam.

Your rolling passes should overlap by three-quarters of the disc diameter or more (Figure 12-64). Remember: this is not an aggressive action, but a fine finishing action. The rolling action with the left hand should go flat to flat through 90 degrees or just a little shy (Figure 12-65).

Figure 12–64 Rolling passes.

Figure 12–65 Rolling action with the left hand.

It's pretty hard to take a picture of the end results. If you did everything right, it should look almost exactly like a formed bend produced on a brake (Figure 12-66).

Paint fill myth. The paint will fill in the scratches. This myth continues to stay alive. The best way to describe why it doesn't work is look at some "Snow Capped Peaks." Just because a little snow is on the mountain, doesn't mean you can't see the mountain (Figure 12-67). Parts show grinding scratches after a light coat of flat paint.

Figure 12–66 The end result.

Paint is a coating that has some thickness, but generally not enough to hide bad grinding. Just ask any auto body shop that does painting how smooth the surface should be prior to painting. The underlying surface needs to be smooth and scratch free. Because most painted surfaces have some gloss to them, they tend to highlight any scratches that were not removed.

Figure 12-68 has the same surface, but I used a finer grit abrasive and a final skim with a non-woven abrasive like Scotchbrite to remove most of the offending deep scratches.

The finish of the finishing chapter has arrived!

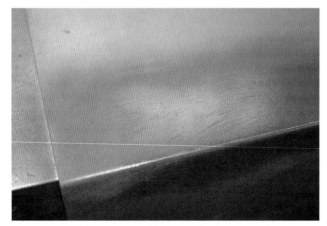

Figure 12–67 Even paint shows scratches.

Figure 12–68 Removing the scratches.

The Junk Drawer

13

13.1 Miscellaneous Tricks Without a Home
13.2 Ideas for the Shop Floor

13.1 Miscellaneous Tricks Without a Home

This chapter is like that special odds and end drawer in your toolbox. You know the one—with all the weird stuff for which you can't find the perfect storage space. Most everybody has at least one. One of the more famous ones is the drawer in your kitchen with all the diabolical food preparation tools that always seem to interlock in ways that make them difficult to remove from the drawer. It's also the drawer where the tool you most want is *always* on the bottom.

When I was a kid my dad had a workshop in the basement. I spent a fair amount of time snooping around because that's what kids do when left to their own devices. There was a special drawer my dad called the top drawer. It was the top-most drawer in a cabinet under the workbench right next to the DC bench grinder my dad got from a navy ship. It was a magical drawer. It had all kinds of strange and interesting things in it. Ever seen a cork boring set? How about the tool to sharpen the cork boring set? Master links, steel balls, three pounds of hex wrenches left over from mounted bearing kits, springs, a veritable mechanical cornucopia. If I was on a deserted island, this would be the drawer I would want with me.

This chapter is kind of like that top drawer. A mix of experiences and techniques that don't fit too well anywhere else, but just like the top drawer they are worth keeping.

By the way, machinists and other metalworkers all have drawers like this. Just for fun, here's a list of items I bet you would find in most machinists' toolboxes or top drawers:

- Steel balls from ball bearings. Machinists can't throw these away, ever.
- A little box of carbide inserts that don't fit any of the tool holders in their box or in the shop.
- Three-to-five insert torx wrenches all the same size. These are way too nice to just throw away.
- At least one broken 6-inch scale.
- A pound or more of dowel pins in assorted sizes and lengths.
- Several pieces of brass rod around 1/4 or 3/8 in diameter, about 3 inches long. Mushroomed ends optional.

If you do any welding work add the following to the list:
- At least one tape measure with the end broken off.
- A ball pein hammer with tape around the handle up near the head.
- A pair of vise grips missing the spring.

Figure 13–1 Fine cast iron shavings.

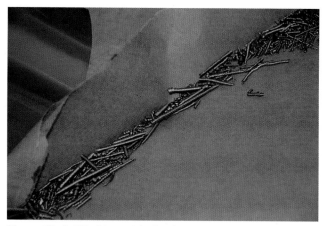

Figure 13–2 Working with shavings.

Shop heat treating. When wrapping parts in stainless wrap for heat treating, add a pinch of fine cast iron shavings to the envelope (Figure 13-1). These burn up during the heat treat and consume the detrimental oxygen inside the wrapping. The small particles of cast iron heat up and burn long before the heavier parts have a chance to scale from the oxygen (Figure 13-2).

Another shop heat treating tidbit is to blow some Argon from the TIG welding torch into the envelope to help exclude oxygen and improve the atmosphere inside the envelope (Figure 13-3).

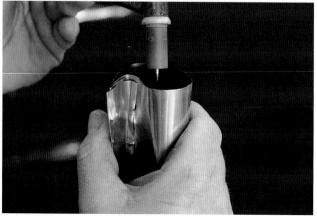

Figure 13–3 Another heat treating tidbit.

Figure 13–4 Working with tiny parts.

Figure 13–5 Advantages of a cleanly swept floor.

Tiny Parts. Sweep the floor before you work with really small parts and assemblies. Those tiny parts resemble chips and debris pretty closely (Figure 13-4). I'm getting too old to crawl around on the floor so I had to get a little smarter. The precious part stands out in stark relief on a cleanly swept floor (Figure 13-5).

Put something in the sink drain before you wash those tiny little parts you just spent three days making (Figure 13-6). I really hate it when I lose a part to plain old stupidity. This beats explaining why you're taking the drain trap apart.

If you have a really diabolical mechanical assembly that is a jack-in-the-box of preloaded balls and springs, do yourself a favor and put the thing inside a clear plastic bag to take it apart (Figure 13-7). Slip your hands inside the bag or do the work from the outside if possible. If you absolutely cannot lose any parts, this is the way to keep them at least contained to a small area.

Figure 13–6 Blocking the sink drain.

Figure 13–7 A mechanical assembly with many parts.

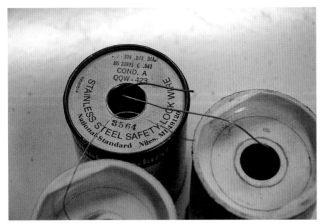

Figure 13–8 Annealed stainless safety wire.

Figure 13–9 Securing small parts.

Safety wire. I love annealed stainless safety wire (Figure 13-8). This stuff is so consistent and pliable, it's like metal taffy. I use it to secure small parts for tricky welding or silver soldering jobs (Figure 13-9).

Stainless safety wire can stand up to the heat of a welding torch without breaking or melting away while still holding your assembly in position (Fig 13-10).

Studs and standoffs. Many times machinists are called upon to cut small lengths of threaded rod to use as studs or connectors in an assembly. Instead of fussing around with cutting and the inevitable

Figure 13–10 Stainless safety wire.

de-burring of all-thread, we buy several popular sizes of long set screws (Figure 13-11). They have the nifty added feature of a small hex driver in the end so you can hold or tighten them. One of our favorite tricks using set screws is in a situation when you need a solid standoff with a male thread (Figure 13-12). We fabricate the standoff with two tapped holes, which is pretty much cream cheese for any shop in any material. Instead of single pointing a male thread, we just install a long set screw into the female thread and—presto!—instant male thread.

Figure 13–11 Long set screws.

Figure 13–12 When needing a solid standoff with a male thread.

Figure 13-13 Tube end forming.

Figure 13–14 Expanding a Hastelloy tube.

13.2 Ideas for the Shop Floor

Tube forming. When in doubt, try it out. Here are a couple of examples of tube end forming that we initially thought were going to be very difficult, but turned out to be very easy. The hexagon was just that—a male mandrel was pressed into the tube with lubricant and then withdrawn (Figure 13-13). The flats were squeezed in a smooth-jawed vise; it fit the hex with a play free slip. The expanded section of this Hastelloy tube was done in small steps with 5C expanding collets (Figure 13-14). Originally the customer wanted the tubes machined from solid. We tried a quick idea with the tubing and made a better part in the end with a lot less frustration. Sometimes all it takes is a willingness to try a wacky idea. Many times with a little curiosity and effort you end up with a new skill and the confidence to use it.

The pivoting fixture was used in the press brake to stretch and straighten small diameter tubing (Figure 13-15). The job was to resize a tube to make a slip fit on the ID of another tube. The matching tube diameter was not commercially available, so we took the next closest size and made a drawing fixture to reduce its diameter. The stretching had the added benefit of making the tube as straight as an arrow. We were able to control the diameter within a couple of thousandths of an inch without trouble.

The way this works is one end of the tubing was anchored to a fixed block off to the left. The opposite end was mounted to this moving block. We used swagelock ferrule fittings to lock the tubing into the fixture. The press brake was fitted with a flattening punch which came down on the roller in the top of the picture. By setting different depths, we were able to control the stretch of the tube very closely.

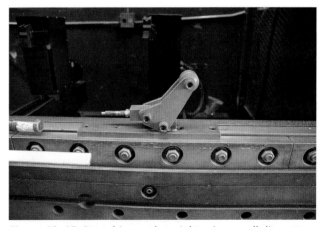

Figure 13–15 Stretching and straightening small diameter tubing.

Figure 13–16 Tube fabrications.

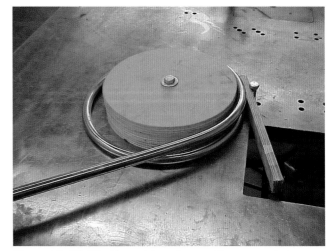

Figure 13–17 Working with a plywood form mandrel.

Tube coiling. Figure 13-16 shows a couple of interesting tube fabrications. They're interesting because we do not have any machinery for coiling tubing. The material is soft copper refrigeration tubing, made with one piece of tubing. The first step was to uncoil the raw material and straighten it for our forming operation. This was 1/2″ outside diameter; it took a little force to straighten it. We used the forklift to assist in the straightening. We secured one end of the tubing to one of our heavy welding tables; the other end we attached to the mast of the forklift. Using the hydraulic tilt feature, we pulled the tubing perfectly straight without any kinks or bumps. We then hand wound the tubing on a plywood mandrel like the one in Figure 13-17. The size of the form mandrel was determined through testing so with the tubing spring back we ended up with the correct size coil.

Figure 13-17 shows the same tube coiling method, but a different project. The plywood form mandrel is just stacked up as high as you need the coil. The plywood mold is bolted to the welding table and a stop wedge inserted to hold the free end.

As in many cases with cutting tools, they can actually be too sharp to cut properly (Figure 13-18). The shape of the broach teeth naturally causes the lower teeth to bite in and tilt the tool. Once the broach tilts, it is cutting the keyway too deep—typically at the bottom where you can't see what's going on. You can carefully stone the edges of a new broach to keep it from digging in (Figure 13-19). Before you jump down my throat and tell me I'm a dimwit, I learned this trick from a guy at Dumont Broach. We were having problems broaching some stainless sleeves, so I called to discuss the problem. It was an enlightening discussion and solved our problem. As a habit, I back off the pressure several times during the broaching operation and check the alignment of the broach to the axis of the work.

Figure 13–18 Cutting tools that are too sharp.

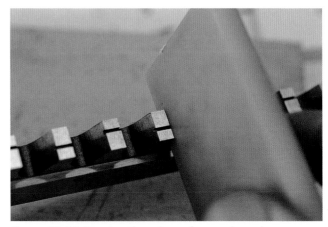

Figure 13–19 Stoning the edges of a new broach.

Figure 13–20 Preventing breakout.

Figure 13–21 Using soft aluminum tape.

Metal tape uses. You can use soft aluminum tape to help prevent breakout in delicate materials (Figure 13-20). The tape is just rigid enough to support those delicate corners. This trick, combined with modifying your tool path so the cut path is toward the interior of the part, can get you out of a sometimes difficult operation (Figure 13-21).

Figure 13-22 shows another great use for soft aluminum tape. This tape will hold up under forming pressure to protect the finish on you materials. It can also be used to prevent iron transfer from the press brake die to a critical work piece.

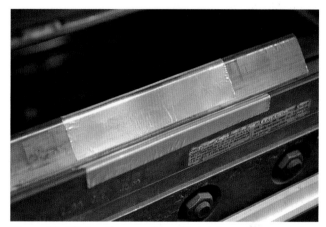

Figure 13–22 Another use for soft aluminum tape.

Machine after heat treat. Try machining some parts after heat treating. 17-4 and 15-5 stainless steel are great candidates. In fact, some of these alloys actually increase in machinability rating with a mid-range heat treat. We have also been able to machine M2 drill blanks directly into cutting tools in a pinch (Figure 13-23 and 13-24). This beats a bunch of tricky setups on the surface grinder any day.

Figure 13–23 Machining M2 drill blanks.

Figure 13–24 An improvement over surface grinders.

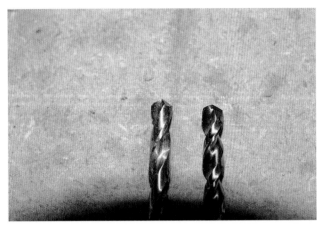

Figure 13–25 Drills with a flatter-included angle.

Figure 13–26 When drilling deep diameter holes.

Deep hole drilling. For small-diameter, deep-hole drilling, use drills with a flatter-included angle 135 or more (Figure 3-25). They drill straighter than steeper angles. Short pecks 1/2 –1 diameter, with a full retract, are the norm with deep 10+ diameter depth holes (Figure 13-26).

Figure 13–27 Cutting gaskets with a hand punch.

Figure 13–28 A tool for straightening and bending small rods.

Gasket cutting. Use the end grain of a block of wood for cutting your gaskets and other soft materials with a hand punch (Figure 13-27). The end grain of the wood exposes the tubular structure of the wood and allows the punch to penetrate cleanly through your material and into the block.

Rod Bending. Here is a little tool I made for straightening and bending small rods (Figure 13-28). It clamps in the bench vise and has a step so the vise can grab it. The pins are plain old dowel pins. This tool is handy if you make hooks and loops out small-diameter rod and wire (Figure 13-29).

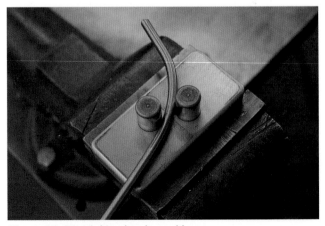

Figure 13–29 Making hooks and loops.

Figure 13–30 A tool for bending steel rod.

Figure 13–31 Bending a loop beyond 180 degrees.

I built the tool in Figure 13-30 many years ago at a place where we had to bend miles of 1/4-diameter steel rod. These bent rods were piped all over the inside of the device; the electricians bundled their wires to the looms to make a neat job of the wiring. The bender clamps in the vise and can bend a loop beyond 180 degrees (Figure 13-31).

Figure 13–32 Another homemade bending tool.

Figure 13–33 The arm pivots and carries a roller.

Figure 13-32 features another homemade bending tool. This one was first built to bend the tubing nerf bars on racing go-karts. It bends 1/2 and 3/4 diameter tubing. As are most things I build, it was made from scrounged leftovers from paying jobs. My version of industrial recycling.

The arm pivots on a bearing and carries a roller with rounded grooves in it to follow the bend (Figure 13-33). The cross pin releases the following roller so you can extract the bent tube from the machine more easily after multiple bends (Figure 13-34).

With the roller follower close, fairly deep offsets are possible (Figure 13-35). When I want to use the bender, I just clamp it to the welding table. I never bothered to bolt it down anywhere.

Figure 13–34 Extracting the bent tube.

Figure 13–35 Making deep offsets.

Closing Thoughts

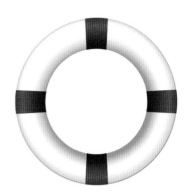

So, once again, we have come to the end of the line — at least for this book. Well, almost — there are still a couple of appendices left! As before, I am happy for actually finishing the book and look forward to a vacation! Also as before, I've found that working on *Doing It Better* helped rekindle my deeper passion for metalworking. Without that excitement about the trade, I could have never finished it.

This was a pretty big project all said and done — in particular, the management of all the pictures needed to bring the book together. I shot something like 15,000 pictures in all, of which only a small portion actually ended up in the book. If you ever need a picture of anything in a metalworking shop, let me know; I might have one.

In the process of writing the original *Sink or Swim* as well as *Doing It Better*, I learned many new things about myself and my trade. I kind of became a metalworking rancher. Let me explain.

A typical rancher has to be skilled in many different careers besides cows and crops. An average rancher ends up becoming proficient at operating and repairing heavy equipment to maintain fields and roads as well as maintain dozens of fossil fuel burning contraptions like pumps and chainsaws. Any decent rancher can negotiate a business contract, buy a hay baler, and not lose his shirt, as well as fell a tree or brand a steer.

What I am getting at is I had to learn about a lot of things I didn't know much about to write these books. Working with the editor and publisher improved my writing and empathy for the reading audience. It also improved my ability to take skills that I learned by doing and explain them in words and writing. Taking the thousands of pictures for this book taught me ways to get better pictures with less effort. My goal has always been to capture as much knowledge about metalworking and get it down on paper so other people can make use of it and have some fun in the process. I hope this book has achieved that to some level.

So if you think you might want to write a book about your skills and life experiences, here is a short list of booby traps I fell into that you might want to avoid.

- A little bit of work on your book frequently is better than a bunch not very often.

- When you think you have taken enough pictures of something you are describing, take three more.

- When you think of something you want to include in your book, quick, quick, write a note so you will remember. I adopted the habit of carrying a small notebook in my wallet for this exact purpose.

Photography

Most of the pictures in this book were shot with a Nikon D70 digital camera. A few of the pictures that appear were older photos I had taken with an older lower resolution camera; many of them I couldn't re-shoot because of the subject matter. I used only two lenses for all the pictures — either a Nikon 18-70mm or a Sigma 1:1 macro.

Photographic lighting was difficult. I used extra lighting only in special occasions. Moving around the shop from one bay to another made setting up fancy lighting quite a bit of extra work, so most were shot with ambient shop light. I did use a tripod when in the shop. Not for every shot, but as many as I could. This is especially important with the macro work.

Organizing the photos would have been nearly impossible without the help of a photo organizing program. There are many out there, but I used Picasa by Google. It was free and had most of the features I needed to organize all the artwork in the book.

I truly hope you enjoyed this book. I certainly had fun and learned a lot writing it. If you have any questions or comments, or would just like to share a great shop story, you can contact me via email at:

- Positive feedback, movie offers, shop tours, etc.   sinkorswimbook@gmail.com
- Negative comments, whining, bellyaching, and redundant questions, etc.   deletemeorsink@gmail.not

Remember: keep learning everything you can about your trade. It's your best protection from early fossilization. Good luck and may your chips be tan and your tools always sharp!

Appendix A
Squaring Blocks without a Tool Change

This appendix summarizes a seven-step method of squaring blocks without a tool change. In particular, this method is useful for squaring up rough sawn blocky parts prior to final machining. There is a particular sequence used to machine all sides of the stock, then produce square and parallel faces using only the bottom cutting surface of a face mill or flycutter.

All that is required are simple tools available in most workshops: a face mill of some sort, a selection of parallels, and a rod perhaps 3/8 (10mm) to 1/2 (12mm) in diameter and 6 inches (150mm) long. Alternatively, a ball bearing with a flat ground on it will work (Figure A-1).

Step 1. The first step is to choose the best surface possible on the rough blank to place against the back of the machine vise. Clamp the chosen surface against the rear fixed jaw of the vise (Figure A-2).

Figure A–1 Simple tools needed for squaring blocks.

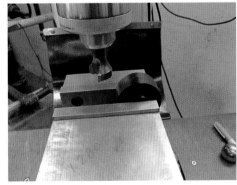

Figure A–2 Clamping surface against the rear fixed jaw.

Figure A–3 Machine the first side.

Figure A–4 Place number 1 surface against rear jaw.

Machine the first side of the rough blank. To help keep track of the cut surfaces I like to number them with a marker (Figure A-3).

Step 2. The next step is to place the freshly machined surface number one against the rear jaw of the vise supported with one parallel only. Use the rod or ball to make sure the surface is bearing only against the rear jaw and the single parallel. The single parallel assures that surface one is the only surface that is being used for registration (Figure A-4). The second surface is now machined (Figure A-5). This second surface in turn is placed against the rear jaw of the vise using the rod to assure the surface is only bearing against the vise jaw (Figure A-6).

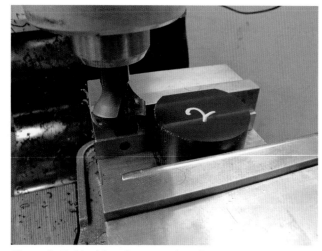

Figure A–5 Machine the second surface.

Figure A–6 Place the surface against the rear jaw.

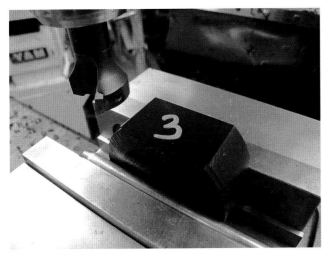

Figure A–7 Placing the third surface.

Figure A–8 Add second parallel.

Step 3. Surface three is now machined flat (Figure A-7) and placed against the rear jaw of the vise (Figure A-8). When surface three is placed against the rear jaw we have two parallel surfaces and a machined bottom surface. At this point we add a second parallel to support the part.

Step 4. Seat the part carefully against the parallels. Make sure the part is sitting down evenly on both parallels. After surface four is machined we have four parallel and, hopefully, perpendicular sides (Figure A-9).

This is the point where it can get a little confusing and why I like to number the surfaces. The part is now clamped at ninety degrees from the previous clamping. Note that this surface number five will be cut twice, so it's not necessary to even put it in the vise squarely. In fact, I purposely put the part out of square to illustrate this fact (Figure A-10).

Figure A–9 The fourth side is machined.

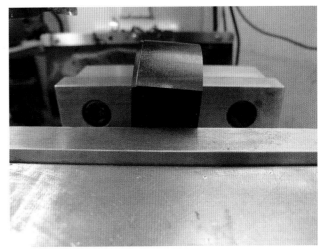

Figure A–10 The part put out of square.

Figure A–11 Machining the fifth surface.

Figure A–12 Placing surface five on a single parallel.

Step 5. When we machine this surface, the only thing that's important is that the two parallel sides against the vise jaws are bearing evenly. I typically machine surface five using the Y-axis. As you will see soon, this produces an edge that is perpendicular as opposed to surface that is perpendicular. We will use the perpendicular edge to register surface six (Figure A-11).

You can see that this surface is not perpendicular to the adjacent surfaces if you look closely. But you will notice that the right hand edge is square to the adjacent face. Just because it's at some arbitrary angle does not mean it can't be perpendicular. This is the key point in this method. I mark this surface carefully so I know it needs to be re-cut.

Surface five is now placed carefully in the vise on a single parallel (Figure A-12).

Step 6. We only want the edge to register on the parallel when surface six is machined. This makes surface six perpendicular to all the surfaces we cut previously (Figure A-13).

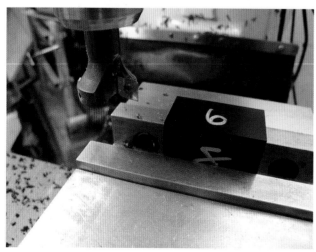

Figure A–13 Machining the sixth surface.

Step 7. The seventh step in the process is to re-cut surface five into final surface seven which we marked carefully so it wouldn't get mixed up with all the other similar looking sides (Figure A-14). This surface will be cut with two parallels under the bottom to assure the cut face is parallel and perpendicular (Figure A-15).

It takes longer to write about how to do this process than it actually takes in the shop. With a little practice with this method you can square up stock quickly and accurately without a tool change. All the surfaces will have the same finish on them and look professional. This process works best with blocky cube-like parts. Thin plate-like parts do not have adequate bearing surfaces to register the surfaces perpendicular.

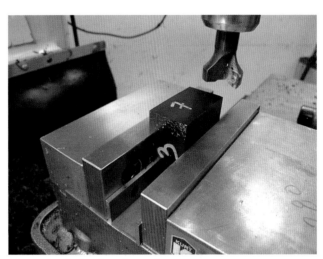

Figure A–15 Cutting the surface.

Figure A–14 Machining the final surface.

Appendix B
Recommended Reading List

Title	Author or Publisher
McMaster Carr Catalog Hard to get but, yes, read it.	McMaster Carr Supply
Machinery's Handbook If you don't have a copy of this book, you aren't a machinist.	Industrial Press
Jorgenson Stock Guide 　　Now called EMJ on the web, you can sometimes weedle one of the stock guides from a sales representative. Otherwise, look for an old copy.	EMJ Metals
Design of Welded Structures 　　Read anything written by Mr. Blodgett.	Omer Blodgett
Design Ideas for Weldments	Lincoln Arc Welding Foundation
Machine Shop Trade Secrets	James Harvey
Metals Handbook V3 Machining	American Society of Metals
Metals Handbook V4 Forming	American Society of Metals
CNC Programming	Peter Smid
Practical Ideas for Metalworking Operations	American Machinist
Morse Tools Machinist Handy Book	Morse Tool Company
Tool Engineers Handbook	McGraw-Hill
Details for Product Design	Greenwood—McGraw-Hill
Illustrated Sourcebook of Mechanical Components	Parmley
Foundations of Mechanical Accuracy	Moore Tool Copany
Holes Contours and Surfaces	Moore Tool Company
Mechanical Drawing	French
Engineer to Win	Carroll Smith
Prepare to Win	Carroll Smith
Mechanisms Linkages and Mechanical Controls	Chironis—McGraw-Hill
Racers Encyclopedia of Metals, Fibers, and Materials	Forbes Aird
How to Run a Lathe	South Bend Lathe
Moving Heavy Things	Jan Adkins
Sheet Metal Pattern Drafting	Daugherty

Internet Resources

http://www.mcmaster.com/
   Hardware, materials, tools. If you can't find what you need here, you have bigger problems.

http://www.metalmeet.com/forum/
   Metalworking forum related to cars

http://groups.google.com/group/rec.crafts.metalworking/topics?hl=en&lnk=sg
   Active metalworking newsgroup

http://www.practicalmachinist.com/vb/index.php
   Active professional metalworking forum

http://www.boedeker.com/
   Detailed plastics information

http://www.onlinemetals.com/
   Small quantity metals supplier

http://www.freemansupply.com/
   Casting and pattern making supplies

http://www.talongripsystems.com
   Work holding

http://www.rjproductsllc.com/rj_vise_soft_jaws.htm
   Soft jaw supplier

http://www.surpluscenter.com/
   Industrial surplus: motors, bearings, and hydraulics

http://www.baileynet.com/
   Industrial hydraulics, bearings, and mechanical tubing

http://www.engineersedge.com/
   Engineering calculators, design information, and conversions

INDEX

17–4 Ph stainless steel   72
1144 steel   73
8620   73

A

abrading   94, 303–317
abrasive paper   307
accuracy   25
Acme lead screws   155
adjustable parallels   159
air conditioning   274
air nozzles   59
air ratchet   170
air supply   58–61
aircraft work, sheet metal   274
aligning   155–156, 190, 213–214
alloys   72–73, 239
aluminum   67, 72
aluminum plate   97, 132
aluminum tape   327
angles   30–33, 158, 228, 241, 278,
   281–282
angular divisions   32
annealing   106
annular cutters   162
ANSI Y14.5   21
anvil   143
arc length   297
arc starting pads   238
arcs   27–28
   lengths   32
area   30
assembly table   56
auger layout   257
automotive work, sheet metal   274
axial shrink   269
axis, isometric   19
axis locking screws   155
axis numbers   279

B

back gage   286
backing plate   133
backstops   156
   screws   214
baffles   301–302
ball hones   308
band saw   96, 101, 105
bands   136
bar stock   62
batteries   91
bead blasting   314-315
bearings   135
belly boards   103
belt sander   311

bench vise   83–84
bench work   83–85
bend allowance   276–277, 281
bending   328–329
bends   278, 299
   change   290–291
   multiple   287
   radius   289
   test   289
berries   87
Bessey sliding bar clamps   234
bevels   227
bins   63
blades   99
blank length   276-282
blanking   296
blanks   129, 194
Blessing, Charlie   5
blocks   90, 101, 333-338
bolts   186
bonding, drawings   22
boring bar   128
boring head   161, 178
boring sprockets   142
boring tapers   134
boron nitride dressing stick   304
bosses   243
bowline   117-119
box and pan brakes   286
braces   253
brake bumping   246–258
brakes   286
breakout details   46
Bridgeport mills   154
broaching   170, 326
bronze wool   309
bronzes   72
burning   3, 229
burrs   99
bushings   129–130, 135

C

CAD   19
calculations   276–282
calibration   81
calipers   22, 142, 159, 226
CAM   209
camber   260, 271
cantilever clamps   91
capping   250-251
carbide inserts   128
carpet protectors   315
carts   57, 62
cast iron shavings   322
casting compounds   166–168, 258
catalogs   68

catching parts   212
C-clamps   188
CCMT insert   127
center height   139–141
center punching   222, 250
centering   139
centering bar   224
centerline   132, 139, 174, 224
ceramic de-burring   315
Cerrobend   168
chamfers   145, 199
changeovers   218
channels   245, 291, 293
chatter   128
chip forming   67
chip loads   74, 212
chords   27–28, 297
chuck jaws   138, 170, 213
chucks   133, 150, 156–157
circles   18, 27–28, 230
circular saws   97
clamps   91, 120, 172, 186–188, 234,
   236–237
cleaning   193
clearance   165, 209
clogging   316
CNC lathe   207–218
CNC mill   181–205
coiling   137, 326
cold rolled steel bars   71
collars   138
collet   129, 192–193
color   66
combination square   156, 224–225
communication   14–16
composite materials   240
computers, cost   35–36
concave radius   178
conductivity   67
cones   297–301
constant surface speed   212
contouring   202–204
convex surfaces   175
cooling   55, 267
copper   58, 84, 140, 237
cordless drill   90
corner reliefs   102
corners   49, 100–101, 110–111, 225,
   281–282, 287–289, 299
cosine   31
counter-bore   135, 170
cracks   223
crib   79–82
cross members   112
crosshatch pattern   308
curvature   246

cutters, annular 162
cutting 99, 110–111, 160–161, 187, 205, 228, 230, 295
   bars 62
   discs 304
   speed 74
cycle times 193–194
cylinder 247–249

D

datum 49
de-burring 199, 309, 315
deep hole drilling 328
density 66
depth of cut 128, 146
details 46
dial thread 149
dial vertical 155
digital drawing 22–23
dimensions 46-49, 291
disconnects 61, 193
discs 234, 306–307, 317–318
dividers 222–223, 282
DOC 128
dolly 57, 121
Dop Wax 197–198
double ended parts 132
dovetail blanks 198
dovetail o-rings 47, 140, 171
drafting 283
drains 323
drain cocks 59
drawbars 155, 157–158
drawing 17–23, 45, 50, 96, 274
dressing stick 304
drill press 309
drilling 90, 159, 169, 174, 328
drilling chucks 156–157
drive dogs 138
Duane, Doug 5

E

ear plugs 76
edges 299–301
elasticity 69
electrical tape 78
electrodes 231–232
electronic drafting 19
electronic work, sheet metal 274
elongation 69
end mill 184, 193
end tabs 48
engineers 44
English wheel 10, 143
engraving 189, 191
equipment, safety 76–78
expanding collet 129
extensions, tap 92

F

fabricating rings 239
face extender 225
face mills 333
face plate 132
feed rate 74
files 147
filing 86–92
filler rods 239, 258, 305
fillets 254
finishing 143, 303
fitup 292, 305
fixture parts 196
flame straightening 259–271
flanges 299–300, 302
flat bar 292, 307, 310
flat blade screwdriver 82
flats 63
flattener 306–307
floors 52–53
flush weld 242
foil tape 168
food areas 55
form tools 130–131
forming 278, 292–293, 325
frames 241, 251
friction cutting 99
frustum development 297
fuel valve 264
fusion tack 296

G

gage 286
gage pins 28, 159
ganged parts 195
Garland rawhide mallets 233
gas lenses 232
gaskets 328
geometry 46
glass beads 314–315
glasses 76, 78
glue 202–204
gooseneck dies 294
grab-stock 197
graining nails 305
grinder holders 306
grinder wheels 313
grinding 106, 112, 130, 234, 303–317
grippers 199
grit 308, 309
grooves 47, 149–150, 231
ground strips 235
guide wheel 229
gussets 238, 292

H

hacksaws 85
hammers 233

hand sketches 17–21, 45
hand tapping 132
handling 62–64
hardhats 76
hardness 67, 279
Hastelloy tube 325
hazards 76
heat 263–264
heat affected zone (HAZ) 93
heat shrinks 263–264, 266, 271
heat treating 322, 327
heating 55, 274
helve hammer 8
hex nuts 142
high-speed cutting 205
hobby shop 2
holdback pin 298
hole saws 163
holes 102, 159, 195, 287, 328
hones 308–309
honing stone 164
hose 193
hose reels 60
hot melt glue 202–204
housings 298
HVAC 274
hydraulic rams 122

I

identifying materials 64–75
indicator 134, 137, 151, 155
industrial work, sheet metal 274
inside radius 279–280, 289–290
interior slugs 184–185
internal threading 146
introduction 1–11
ISO ruler 18
isometric axes 19
isometric circle 18
isometric projections 17–21

J

jacking screw 169
jaw covers 140
jaw spacer 109
jaws 170
joints 244–245
journeyman 7

K

keyways 170

L

laps and lapping 313
laser drawing 22
lathe
   center height 139–141

CNC 207–218
    long stock guide 141–143
    manual 125–151
    operators 49
    part making 212–218
    programs 209–211
    step turning 144–147
    threading 144–147
    tooling 128–129
    workholding 129–135
layout 222–232, 258. 275–276
layout lines 113
legs 237
length 71, 228
lenses 232
levels 120, 173, 227
levers 123
lifting 116–123
light 54
lines 27
logbooks 196
long stock guide 141–143
lugs 119

M

machine oil 308
magnetism 66
magnets 66, 82
mallets 233
mandrel 326
manual lathe 125–151
    center height 139–141
    long stock guide 141–143
    step turning 143–144
    threading 144–147
    tooling 128–129
    workholding 129–135
manual milling machine 153–179
mapping 264–265
masking plate 312, 314
mass 30
materials 65–75, 278
    composite 240
math 26–29
Maudslay, Henry 125
measuring tools 227
metal tape 327
metalworkers 44
metric system 33–34
micrometer 147
mill parts 160–161
milling,manual 153–179
miter saws 98
mitered corners 110–111, 282, 299
modulus of elasticity 69
mosaic drawing 22
multiple cuts 109
multiple start threads 147–151

N

neutral axis numbers 279
Niles machine 125–126
Norbide 304
notches 299
notching patterns 281–282
nozzles 59
nuts 146

O

oil canning 287
ordinate dimensioning 48, 49
orientation 49, 129
o-rings 47, 140, 171, 185–186

P

pads 238, 255
paint 320
parallel rolls 226
parallels 19, 275–276, 333
part holding 200–201
part making 212–218
parts 171–172
patterns 283
peening 94, 288
perforated sheets 294
perpendicular strips 247
physical templates 15
PI tapes 226
pipes 58, 72
plastic tips 74
plastics 75, 315
plates 69–71, 133, 188, 240, 259
plywood 326
pneumatic grinders 234
pocket screwdrivers 66
point of tangency 27
polishing 310, 318
polycarbonate sheet 314
porous castings 258
power notcher 281
power rolls 298
pre-camfering 211
precision balls 163
pre-clamping 140
pre-heating 257
press brakes 288
Pro-Cast 167
profiling 186–188, 197–198
projections, isometric 17–21
proofing 216–217
protection 95
prototypes 197
protractor 158, 227
prying 123
pulling 291
punch marks 250

push sticks 104
pushing 291
PVC 58, 70
pyramid rolls 10, 298

Q

quick changes 217
quick vise grip 137
quill locking 155

R

racks 63
radial patterns 48
radial shrink 269
radius 28, 88, 143
radius formula 290
radius grinding 319-320
raw materials 62–64
recessed scale 98
rechargeable batteries 91
reference bar 287
relief hole 287
repetitive cuts 295
retractable point scriber 82
rigging 116–123
right angles 161
right hand left hand drawings 50
right triangle 178
rigid tapping 202
rims 299–301
rings 239
rods 3, 107, 239
rod bending 328–329
roll grinding 319–320
roll method 284–286
ropes 117–119
rotary table 161, 175
rough cut blanks 129
rough setting 292
roughing 16
rounding 48, 284
rounds 101, 195, 256
ruler 81
rules 226
runout 192

S

safe edge 86
safety 76–78
safety wire 324
sanding 223, 303–317
sanding discs 306–307, 317–318
saw blades 99
saw horses 237
saws 93–115, 163
scale 98
scalpel 315

score marks 127
Scotchbrite sheets 305
screwdrivers 66, 82
screws 144, 155, 214, 324
scribe lines 82, 223
shaft collars 138
shafts 259, 268–269
Sharpie 82, 139, 275, 282
shavings 322
shearing 296
sheet metal 273-302
shop math 26–29
short wrench 168
shrinkage 254, 263, 267, 269
sine one degree 32
sine 31
sine bars 158–159
sine plate 173
sink drain 323
sketching 17–23
slings 119
slugs 162, 184-185
small rods 107
snipping 287
soft center 139
soft jaws 142, 189, 210
Sohcahtoa 31
spacer 109
speed 25
speed tools 92
spherical surface generation 174–179
spider 136, 142
spin handle 138
spindles 74, 173, 215
split parts 114–115
split sleeve 134
spreader bars 164
spring hand clamps 283
springs 137
sprockets 142
square corners 100–101
square root of 2 32
squares and squaring 134, 245, 253, 255, 333–338
stability 84
stacked parts 171
stainless polishing 310
stainless steel 72
standoffs 324
steel 58, 71, 72–73
step boring 128
step turning 143–144
stock 45, 62
storage 62–64
straightening 259–271
strap clamps 172
stress-proof steel 73
structural beams 241

studs 324
stumps 108
sub-assemblies 242
sub-containers 316
sub-plate 172, 187
supplies 59–60, 68
Surface Armor 95
surface generation 174–179
surfaces 255
swivel bases 83

T

tables 56–57, 120, 235–245
tack welding 237, 253
tacking 296
tailstock 136, 140, 151
tangency 27, 31
tanks 301–302
tap extensions 92
tap guide 90
tap wrenches 91
tape measure 81
tapers 134, 292
tapping 159, 169, 191, 202
tea strainer 314
temper 278
templates 15, 275
tension block 137
test indicators 137
theoretical sharp 46–47
thermal conductivity 67
thickness 71, 274, 278
thread dial 149
thread files 147
thread inserts 243
threading 144–147, 191
three-dimensional sawing 102
three-jaw chuck 132, 205
three-point contact 84, 120, 163
TIG torch and welding 93–94, 230, 322
tiling 23
tipping heads 172–174
tolerances 45, 221, 277
tool bits 130–131
tool holders 130
tool marks 314
tool runout 192
tools 79–82, 128–129, 213–214, 227, 233–234, 278
toughness 69
trammels 222, 227, 282–283
tramming 155–156
traversing 145
trigonometry 30–33
trucker's hitch 117–119
true bands 136
tube coiling 326
tube corners 225

tube ends 250–251, 325
tube fitup 242,305
tube forming 325
tubes 226, 261, 268–269
tubing 245
tumblers 316–317
tungsten 231–232
turning 138–139
tweezers 78
twist drill 170
twists 252

V

Van Bebber, Fred 5
V-blocks 256, 270
Vee-type dies 279, 289
ventilation 274
vertical band saw 93
vertical dial 155
vertical milling machine 154
vibratory tumblers 316–317
vise 83–84, 105, 137
vise jaws 170
volume 30

W

warped discs 306–307
washers 239
WD-40 217, 308
wedges 121–122, 156, 316
weld joints 244-245
welding 47, 93–94, 106, 108, 201, 219–258
    areas 52
    fillet 254
    layout 222–232
    table 235–245
    tolerance 221
weldments 271
wheel, English 10
whip 60
Whitworth, Joseph 125
widow punch 295
wire 252
wire brush 312
wire tension block 137
WNMG insert 127
workbenches 56-57

Y

"yank" method 284–286
yield strength 69

Z

Z-axis 132